T0135717

Mathematical modelling of molecular networks in cancer cells using modular response analysis

Bertram Klinger

Dissertation
Humboldt-Universität zu Berlin
Lebenswissenschaftliche Fakultät
2016

Bibliografische Information der Deutschen Nationalbibliothek

Die Deutsche Nationalbibliothek verzeichnet diese Publikation in der Deutschen Nationalbibliografie; detaillierte bibliografische Daten sind im Internet über http://dnb.d-nb.de abrufbar.

ISBN 978-3-8325-4299-3

Logos Verlag Berlin GmbH
Comeniushof, Gubener Str. 47,
10243 Berlin
Tel.: +49 (0)30 42 85 10 90
Fax: +49 (0)30 42 85 10 92
INTERNET: http://www.logos-verlag.de

Contents

Introduction

Nature is relentless and unchangeable, and it is indifferent as to whether its hidden reasons and actions are understandable to man or not

— Galileo Galilei (1564 - 1641)

It is an inherent property of scientific research to perpetually search for means to push the frontier of the unknown yet a little further. The discoveries that were made in that process represent most often merely tiny steps which go unnoticed by the general public. However, some of them challenged and fewer even changed the world-view of whole civilisations. The invention of the telescope in 1608 undoubtedly stands for one of those exceptional leaps. With the development of the "first extension of one of man's sense" (van Helden, 1977), a golden age of astronomic discoveries was triggered, headed by the above quoted Galileo Galilei as one of its pioneers. Soon after, Galilei and others realised that next to observing far away objects a similar construct could be build to magnify close-by objects as well (Singer, 1914; van Helden, 1977). Intriguingly, it turned out that the world that opened up in front of those microscopes appeared to be equally vast in depth as the sky seen through a telescope. When exploring the microcosmos with more and more advanced instruments many secrets of nature could be learned. For example it was discovered that the body of all living beings is composed of structurally separated entities which were named cells. Even more surprisingly it turned out that all cells originated from other cells may they be single- or multi cell organisms (Mazzarello, 1999). It became clear that the secret of life itself must be hidden inside those cells. This gave rise to many new disciplines in biological research studying the plethora of molecules that are harboured by the cells, most prominently biochemistry, genetics, and molecular biology. The ensuing research in the past two centuries has helped to identify and functionally assign molecules to live-defining processes such as metabolism, heredity, environment interaction, proliferation and development. It was further deduced that there has to exist a regulatory network that tightly controls those processes. As the molecular identification of the regulatory network members is nearing completion for some model organisms (Müller and Grossniklaus, 2010), the current focus is shifted to the illumination of the regulatory functions and inter-molecule interactions of which, by now, only a small fraction is described. For this task the field of molecular systems biology has been formed at the beginning of the 21st century (Ideker et al., 2001; Kitano, 2002).

In the past decade advanced high-throughput technologies allowed to simultaneously measure many molecules under various experimental settings which has helped to better describe regulatory relationships. Interestingly the more research is conducted to unveil the network wiring, the fuzzier the network appears. This might be an inherent property of the network itself enabling it to integrate many diverse signals. Partly, those contradicting observations can be related to temporal or spatial separation. For example, short term activation of ERK, a growth-signal activated protein kinase, induces growth, whereas long term activation triggers differentiation (Marshall, 1995). Other discrepancies might arise from shared effectors of different pathways. One of the most versatile crosstalk hubs can be found in the immune response-associated transcription factor NFκB on which many signalling pathways converge (Oeckinghaus et al., 2011). Despite the rapid progress in technology, our current abilities in capturing the activation state and interactions of signalling molecules in time and space are limited. Yet, next to curiosity there exists a pressing need to understand the intra- and intercellular signalling, as diseases, in one way or another, all exploit these systems. The arsenal ranges from a few specific processes - abused by many hit-and-run pathogens to enter cells, evade the immune system or usurp the host's metabolism - to a highly complex illness that emerges from the inside: cancer. To tackle this versatile foe, it is of immense importance to thoroughly understand the changes in the regulatory network that have lead to the malfunctioning.

In the effort of unveiling and analysing those complex networks, mathematical modelling has become a vital companion in the field of systems biology (Kitano, 2002; Kholodenko et al., 2012). In the past it has proven successful to adapt well-tried concepts and mathematical frameworks from various research fields. It can be assumed that by now, the majority of feasible methods and concepts have been introduced into systems biology. Therefore the modellers task is to sharpen the existing arsenal of methods and to select the right combination of tools to solve a particular problem in biological science - just as a shrewd combination of a concave and a convex lense, originally made for mere spectacles, enabled the construction of telescopes and microscopes (van Helden, 1977).

In this spirit the present thesis attempts to apply the mathematical method Modular Response Analysis (MRA) to biological data and investigates its ability to reverse engineer and analyse genetic and signalling networks. MRA is itself a good example for the extension of existing principles as it has been developed from the theory of metabolic control analysis (MCA) designed to model metabolic fluxes (Bruggeman et al., 2002). In this work MRA will be tested and adapted to analyse various molecular networks known to play a major role in cancer biology.

1.1 Challenges in targeted cancer therapy

Cancer is one of the world's most leading causes of death accounting for about 25% of all fatalities in Germany (Destasis, 2013) and the United States (Siegel et al., 2013). It is a disease that originates from endogenous cells that by genomic alterations gained the ability to escape the innate growth control (Hanahan and Weinberg, 2000). Cancer - although ever present - has become a more imminent threat when mankind started moving towards a post-infectious-disease era. Accordingly between 1900 and 2010 the US cancer death rate has tripled (cf. Fig. 1.1) - a time period in which the death rate for almost all other major diseases has declined (Jones et al., 2012). The first treating options that have been established were targeting the most prominent feature of cancer cells: excessive growth. Thus the classical treatment strategy involves, next to surgical removal of accessible solid tumours, chemotherapy (sublethal doses of cytotoxins) or radiotherapy both aiming to damage the rapidly dividing cells and thereby many cancer cells. However,

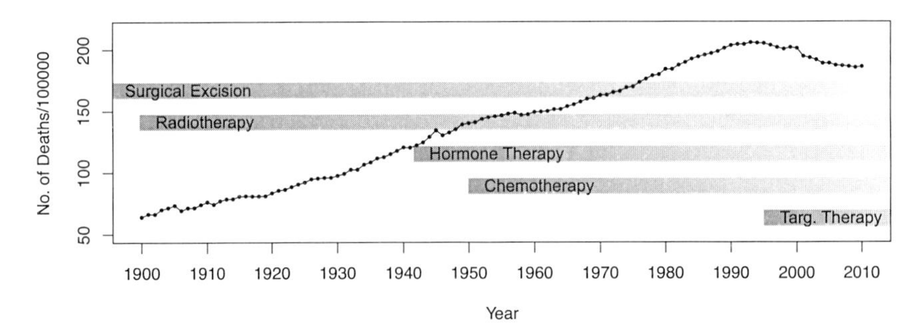

Figure 1.1 Development of cancer death rate and treatment strategies Timeline of US cancer death rate from 1900 - 2010 shows a steady increase until 1993; *data taken from (Jones et al., 2012).* To illustrate the clinical options to treat cancer patients at each particular time point the major therapies are shown as vertical bars according to their first application; cf. Mukherjee (2011). The recent modest decline in cancer death rate can be attributed to earlier detection, improved treatment options and better prevention measures (Jemal et al., 2010; Siegel et al., 2013).

the side-effects of the latter treatments were severe and the healing chances, especially at later progression stages, remained unsatisfactorily low. It was only after the cause and molecular origin of cancer was found that those treatments could be improved and other, more specific, treatments be devised.

A major milestone in cancer research was the discovery that malignant transformations are caused by specific mutations in the genome, called driver mutations (Haber and Settleman, 2007). Current estimates found a driver potential in about 140 genes of which a typical cancer cell bears mutations in 2 to 8 of those genes (Vogelstein et al., 2013). Genes affected by driver mutations can be classified into (i) tumour suppressors that become inactivated, such as p53, the "guardian of the genome", and (ii) oncogenes that cause permanent signalling like the growth signal transducer RAS, a small GTPase connecting growth receptors with downstream effectors. More intriguingly, it was found that it often suffices to revert only one of those driver mutations to stop tumour cells from growing (Weinstein, 2002; Weinstein and Joe, 2008); triggering the advent of targeted therapy. Since the reversion of an inactivation is often more complicated than the inhibition of an overactive signal, most efforts were put into the development of inhibitors of oncogenes and their downstream pathways. The observed addiction of the tumour to certain signals has been noted earlier for certain hormone-dependent cancers, such as prostate or breast cancer (Huggins and Stevens, 1941; Cole et al., 1971). With the newly discovered oncogenes a much wider range of tumours could potentially be targeted.

As of now, two branches of targeted therapy have been established, (i) monoclonal antibody therapy and (ii) small molecule kinase inhibitor treatments. The first successful antibody therapy, was developed against the growth receptor HER-2, that is overexpressed in 20-30% of breast cancers (Hudziak et al., 1989). This approach is highly specific but can only target extracellular and transmembrane proteins such as receptors, ligands or tumour stroma proteins (Adams and Weiner, 2005). Small molecule kinase inhibitors on the other hand can enter the cells but are often less specific and therefore prone to off-target-effects. The first clinically approved inhibitor, imatinib, was derived from an extensive screen to target the abnormal fusion protein of the two kinases Bcr and Abl that occurs in 90-95% of chronic myelogenous leukemia cancers (Shtivelman et al., 1985; Druker et al., 1996). Its success ensued the development of small molecule inhibitors against other hyperactive kinases (e.g. BRAF (Tsai et al., 2008)) and overexpressed receptors (e.g. EGFR (Moyer et al., 1997)).

Treating patients with targeted therapy is, however, only beneficial if the mutational background permits it. It is generally assumed that inhibition of a receptor should not be effective if an activatory mutation is present downstream (Karapetis et al., 2008; Amado et al., 2008).[1] Therefore biomarkers for sensitivity or resistance are searched after to ensure the correct treatment mix for the patients, thus moving a step closer to personalised cancer therapy.

However, even after correct stratification of patients, targeted therapy often becomes ineffective over time, resulting in eventual relapse (Piccart, 2008; Poulikakos et al., 2011; Diaz et al., 2012; Misale et al., 2012; Holohan et al., 2013). Furthermore, for many patients no targeted therapy is currently available. For example, we are lacking a potent inhibitor for RAS, one of the most frequently mutated genes in solid cancer (Roberts and Der, 2007; Kandoth et al., 2013). In addition, only a small fraction of existing signalling inhibitors that target crucial mediators have performed well in clinical trials. For example inhibitors of MEK a Map-2-kinase downstream of RAS performed poorly in cells containing RAS mutations but proved effective in BRAF mutants - which is one signalling step below (Solit et al., 2006). This might be due to the various feedbacks acting on virtually all signalling levels that confer robustness and temporal flexibility to the cell (Legewie et al., 2008; Blüthgen and Legewie, 2008).

To address the aforementioned problems and hence to enable the development of more efficient cancer treatments, it is essential to more thoroughly understand the relevant cell biology. Due to the versatility of cancer, this encompasses a large proportion of the regulatory network in cells (Hanahan and Weinberg, 2011). By better understanding the wiring we may be able to find more predictive biomarkers and also where to interfere best to efficiently eliminate the tumour cells with less side-effects and smaller chances of relapse.

A very powerful measure of our understanding of network relationships is to place this knowledge into a model and analyse its predictability. In this manner the limits of current understanding can be pinpointed, giving the appropriate direction for follow-up experiments. For this task computational systems biology is required to guide the scientist through the maze of available noisy data and apparently flexible biology of the cell (Kreeger and Lauffenburger, 2010).

1.2 Modelling in systems biology

In the past 20 years the focus of biological research has shifted from single molecule research to how molecules are organised into networks (Kholodenko et al., 2012). The networks are traditionally classified into three types: (i) signal transduc-

[1]This view will be challenged by the results of chapter 3

tion networks that sense and relay internal and external cues, (ii) gene regulatory networks that orchestrate the expression levels of mRNA and proteins and (iii) metabolic networks describing anabolic and catabolic processes. Although all three types contribute to the hallmarks of cancer (Hanahan and Weinberg, 2011), the focus of this work was laid on the former two as metabolic networks require a fundamentally different modelling approach (Kholodenko et al., 2012).

The task that mathematical and computational models have to fulfil in systems biology is the mechanistic interpretation of complex observations which elude an intuitive explanation. Take as an example a linear cascade of three kinases which contains a single feedback from node three to node one. Depending on the strength and sign of the feedback this can result in an entirely different kinetic of the terminal kinase. Negative feedback scenarios can produce transient peaks, damped or sustained oscillations whereas positive feedbacks can even bring about bistability (Kholodenko et al., 2010). In the corresponding biological antetype, the RAF-MEK-ERK cascade, this is further complicated by the action of feedbacks on different signalling levels (e.g receptors and downstream mediators) acting on various time scales, e.g. fast posttranslational and slower transcriptional feedbacks (Kolch et al., 2005). The currently most accepted explanations for the existence of feedbacks are to confer robustness to the network response (Blüthgen, 2010; Fritsche-Guenther et al., 2011), to shape the dynamics of the signal and to integrate different signals that utilise the same pathway (Shankaran and Wiley, 2010). Due to these functions, feedbacks are expected to be an omnipresent phenomenon in both signalling as well as genetic networks, that can hardly be correctly interpreted without the help of mathematical modelling.

Often, from an experimentalist point of view the distinction between modelling in systems biology and bioinformatics is not clearly drawn as they are both seen as 'tools' for data handling. While bioinformatics is employed for the important task to statistically process and analyse (hight-throughput) data, systems biology seeks to combine statistical and mathematical concepts with an underlying rationale that strives to go beyond mere description. In the past, systems biology approaches have been successfully applied to a range of tasks including simulation and structural analysis of complex networks (Schoeberl et al., 2002; Schulthess and Blüthgen, 2011), model selection to mechanistically explain experimental observations (Sturm et al., 2010), representation of complex data in a more accessible way (Lau et al., 2011), data-supported network analysis to identify key members and processes (Sahin et al., 2009; Heiser et al., 2009; Cirit et al., 2010), and reverse engineering of network wiring (de la Fuente et al., 2004; Nelander et al., 2008; Kholodenko et al., 2002; Geier et al., 2007). In these approaches the applied techniques and input requirements vary tremendously. Therefore, in order to choose the right modelling approach, it is vital that the question that modelling should answer is thoroughly formulated in advance.

In this work the focus is put on further developing a method that can reverse engineer directed networks from minimal datasets and is still capable of modelling networks quantitatively, including feedback loops. In order to chose the right modelling base the type of input data from which the network will be build has to be determined first. The most informative way to infer molecular networks is to directly study interactions and bindings. However, as these data are not easily derivable from living cells, network inference is often forced to rely on observations of the outcomes of the interactions instead, i.e. altered gene expression or phosphorylation state. The best data type to infer causative networks from these indirect measurements is to observe the impact of well-defined perturbations realised by means of RNA interference, overexpression or kinase inhibitions. These data provide network information, are experimentally feasible and are already widely available in public data bases such as Gene Expression Omnibus (GEO) [2] (Barrett et al., 2013).

1.3 Perturbation data-based models

Inferring regulatory relationships between genes or signalling molecules from perturbation data is one of the big challenges in computational biology. The major problem in disentangling the network structure is that a perturbation of a single node results in direct and indirect effects on other nodes. For example a transcription factor knockdown results in expression changes of direct targets among which other transcriptional regulators are encoded that propagate the signal further. The problem arises if within this transcriptional cascade downstream regulators have target genes that are also regulated by an upstream or parallel regulator.

In order to discriminate cause from correlation and to comprise feedbacks it is vital to generate data that allow the construction of directed regulatory networks. Two data collection strategies have shown this potential: (i) time series data and (ii) pre- and post-perturbation steady state measurements.

Reverse engineering from time series data has been successfully applied by the use of e.g. ordinary differential equations (ODE) (Barenco et al., 2006; Stark et al., 2003) or dynamic Bayesian networks (Geier et al., 2007). Here, causality of regu-

[2]http://www.ncbi.nlm.nih.gov/geo/

lation is inferred by the time delay between the expression of regulator and targets which requires an appropriate sampling rate. Since the optimal sampling rate is not known, a substantial number of time points has to be measured for each perturbation, rendering this approach experimentally laborious for many perturbations and larger networks.

In contrast, steady state measurement-based methods need a larger number of perturbations to infer causality. However, since for each perturbation only two time points are measured, smaller overall data sets are required. Common modelling approaches for this data type are utilising Boolean or Bayesian models (Saez-Rodriguez et al., 2009; Morris et al., 2011), or model the system based on the view of the network as a dynamical system (Gardner et al., 2003; Santos et al., 2007). However, only the latter approach is capable to incorporate feedbacks. Thus, modelling steady state data with the dynamic network approach might best handle the trade-off between data quantity and network information.

1.3.1 Dynamic network modelling

In a dynamical network system the change of the nodes x_i can be described by a set of ordinary differential equations:

$$\frac{d\mathbf{x}}{dt} = \mathbf{f}(\mathbf{x}, t) \quad . \tag{1.1}$$

Most approaches then apply a linearisation of the underlying dynamical system around the (unperturbed) steady state value (Stark et al., 2003; Brazhnik, 2005):

$$\frac{d\Delta\mathbf{x}}{dt} = \mathbf{J}\Delta\mathbf{x} \quad \text{with} \quad J_{ij} = \frac{\partial f(x)_i}{\partial x_j} \quad , \tag{1.2}$$

with $\Delta\mathbf{x}$ representing small deviations in the system variables around the steady state. J_{ij} are the elements of the Jacobian matrix $\mathbf{J}$ and represent the direct influence from variable j to variable i. These entries are zero for no direct connection and, if non-zero, the sign indicates the type of influence, where negative and positive values represent inhibition and activation, respectively. Therefore, sign($\mathbf{J}$) can be interpreted as an adjacency matrix of the network and the values of J_{ij} reflect the strength of regulation.

Persisting perturbations can formally be introduced into the system by adding a vector $\mathbf{u}_j$ to the right hand side of Eq. 1.2. The steady-state response after perturbation can thus be formulated as :

$$0 = \mathbf{J}(\Delta\mathbf{x}_j)_{ss} + \mathbf{u}_j \quad . \tag{1.3}$$

This is the central equation to be solved in reverse engineering. If the perturbation affects only node i, the entries of vector $\mathbf{u}_j$ are zero except for the entry of the perturbed node. Under most settings, the identity of the perturbed node is known but often the initial perturbation strength cannot be quantified. The entries of the vector $(\Delta\mathbf{x}_j)_{ss}$ correspond to the steady state measurements to perturbation $\mathbf{u}_j$. For multiple perturbation experiments, the equations can be arranged in matrix form. Then, $(\Delta\mathbf{x})_{ss}$ represents a matrix where all measurements for one perturbation are stored column-wise.

Attempts to apply Eq. 1.3 to infer networks from steady state perturbation data resulted in two successful approaches (Camacho et al., 2007). The first, called Network Identification by multiple Regression (NIR) was introduced by Gardner et al. (2003) and searches for a $\mathbf{J}$ enforced for sparsity (see Suppl. Section A 4 for a detailed description). The second approach is based on estimating the Jacobian matrix in its entirety. This method has been independently proposed in two flavours: One flavour, called Regulatory Strength Analysis, aims at estimating the Jacobian itself (de la Fuente et al., 2004). The other, called Modular Response Analysis (MRA), estimates the dimensionless, scaled and thus more robust version of the Jacobian matrix (Bruggeman et al., 2002; Kholodenko et al., 2002; Andrec et al., 2005). The disadvantage of NIR compared to MRA is that it requires the knowledge not only of the identity but also the initial strength of the perturbation, which requires additional experiments. Therefore, MRA emerged as the most practicable choice.

1.3.2 Modular Response Analysis (MRA)

Modular response analysis is an approach that extends parts of metabolic control analysis (MCA) (Kacser and Burns, 1973; Heinrich and Rapoport, 1974) to non-metabolic networks such as signalling and genetic networks. The method focusses especially on the concept of response coefficients (Bruggeman et al., 2002). The response coefficients are categorised into two classes: (i) global control coefficients R^i_j and (ii) local response coefficients r^i_j (Kholodenko et al., 1997).[3] R^i_j can be interpreted as the global steady state response of the network on node i

[3]Notation according to Bruggeman et al. (2002); lower and upper indices indicate column and row, respectively.

to a perturbation acting on node j, thus showing the result of direct and indirect interactions. On the other hand r^i_j stands for the steady state response of node i if all other interactions except the one between node j and i are severed, i.e. it encodes only the direct interaction strength.

These response coefficients represent normalised variables of Eq. 1.3 and are defined as:

$$R^i_j := \frac{d\ln x_i}{d\ln x_j} \approx \frac{\Delta x_{ij}}{x_{ij}}$$
$$r^i_j := -\frac{J_{ij}}{J_{ii}} \quad .$$

Entries of matrix $\mathbf{R}$ can be experimentally estimated. In contrast to this, the entries of $\mathbf{r}$ cannot be derived experimentally from intact cell systems. They compose the network structure and their derivation from global measurements is the main challenge for network reverse engineering.

The relationship of the two response coefficients is described in the central equation of MRA. Assuming the diagonal matrix diag($\mathbf{p}$) represents the quantification of single perturbations, Eq. 1.3 can be reformulated as:

$$0 = \mathbf{rR} + \mathrm{diag}(\mathbf{p}) \quad .^4 \tag{1.4}$$

According to Kholodenko et al. (2002) $\mathbf{p}$ can be written as a function of $\mathbf{R}$, which further simplifies the equation to:

$$\mathbf{r} = -\mathrm{diag}(\mathbf{p}) \cdot \mathbf{R}^{-1} = -[\mathrm{diag}(\mathbf{R}^{-1})]^{-1} \cdot \mathbf{R}^{-1} \quad . \tag{1.5}$$

Therefore $\mathbf{r}$ can be directly derived from the global response matrix alone.

The resulting workflow to derive $\mathbf{r}$ in a top-down approach is exemplified in Fig. 1.2. Given a network of n nodes which can be individually perturbed (e.g. by knockdown), measure the systems initial steady states $\mathbf{x}_{(0)} = (x_{(0)_i})_{i=1...n}$ and the steady states after n single perturbations $\mathbf{x}_{(1)} = (x_{(1)_{ij}})_{i,j=1...n}$. Then, the entries of the global response matrix can be calculated by its practical definition (Kholodenko et al., 2002):

$$R^i_j := \Delta_j \ln x_i = \ln\left(\frac{x_{(1)_{ij}}}{x_{(0)_i}}\right) \approx \left(\frac{x_{(1)_{ij}} - x_{(0)_i}}{< x_{(1)_{ij}}, x_{(0)_i} >}\right) \quad . \tag{1.6}$$

[4] Response coefficient transformation requires perturbation u_j to be adjusted with respect to the chosen realisation of R^i_j, indicated by the use of p from now.

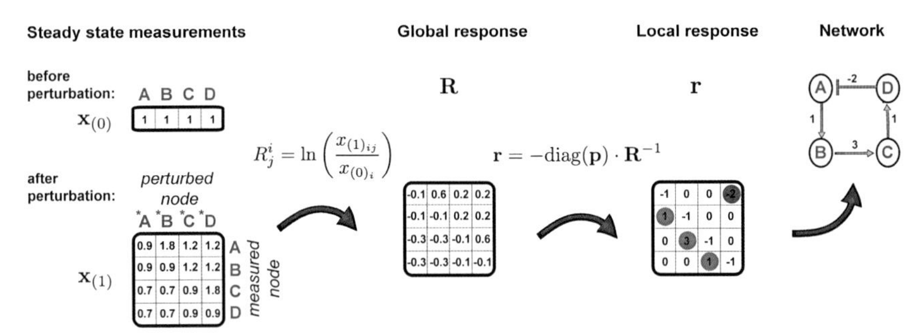

Figure 1.2 The basic principle of Modular Response Analysis Workflow of MRA-mediated network reconstruction from systematic perturbation data. As data input MRA requires the steady state data of all nodes before perturbation ($\mathbf{x}_{(0)}$) and after single node perturbation ($\mathbf{x}_{(1)}$). This information is then combined to derive the global response matrix $\mathbf{R}$. It can be noted that in here $\mathbf{R}$ is a full matrix, stating that every perturbation reaches all other nodes. By applying the central equation of MRA to the global response matrix, the local response matrix $\mathbf{r}$ can be determined, which contains the network structure. In the example an initial perturbation strength of -50% was taken, which for the chosen realisation of $\mathbf{R}$ is translated into the entries of $\mathbf{p}$ as $\ln\left(\frac{1}{2}\right)$.

Through determination of the local response matrix from the global response the local network connectivity along with its connection strength can be unveiled. In the illustrated 4-node example it is shown how a sparsely connected network can be represented by a full global response matrix, rendering an intuitive derivation impossible. This particular effect is due to the existence of a feedback from D to A which allows each perturbation to reach all nodes in the network.

As a consequence of the local response coefficient being derived from the difference after global perturbations, its quantification does not represent a kinetic parameter but the relative influence that the incoming node exerts on the outgoing node. Such for $r^i_j = 1$, a change in node j leads to the same (relative) change in node i. In this manner the value of local response coefficients classifies into five qualitatively different relationships (Kholodenko et al., 1997):

$$
\begin{aligned}
r^i_j > 1 &: \quad \text{Amplification} \\
r^i_j = 1 &: \quad \text{Neutral relay} \\
1 > r^i_j > 0 &: \quad \text{Attenuation} \\
r^i_j < 0 &: \quad \text{Inhibition} \\
r^i_j = 0 &: \quad \text{No connection}
\end{aligned}
$$

Note that the local response coefficient-derived network is not hard-wired (Kholodenko et al., 2012) since the connection strength can vary depending on perturbation strength, time point and cell system.

An advantage of the MRA concept is the generous definition of nodes. They are not confined to represent single reaction molecules but modules that are defined by an incoming and outgoing signal as long as linear approximations can be justified (Kholodenko et al., 1997). Therefore if not all nodes can be perturbed or measured MRA allows to condense nodes into modules until the remaining nodes fulfil the systematic perturbation criteria. In such a way all hidden processes within the module are summed up to one output. Hence the name Modular Response Analysis.

1.4 Obstacles for field application of MRA

With its ability to quantitatively model cyclic networks from mere perturbation data, MRA is in principle a powerful tool for many existing data sets. It should be noted, however, that the MRA approach in its original definition is not applicable to experimental data for various reasons:

noise Experimental data are noisy and thus the global response matrix has to be derived from noisy data. A noisy $\mathbf{R}$ will propagate the noise to the local response matrix, where all entries will be different from zero pretending a fully connected network. In addition, through inversion of $\mathbf{R}$ noise will be amplified unequally. Thus, simply thresholding the entries of $\mathbf{r}$ will not be sufficient to distinguish between spurious and real interactions. Likewise the estimation of the primary perturbation $\mathbf{p}$ from the global response matrix in Equation (1.5) is hampered.

linear approximation Furthermore MRA assumes a linearised system, however, many biological reactions are not strictly linear (e.g. a double phosphorylation event), which is aggravated when modularisation is introduced. Although it can be assumed that due to steady state observations many non-linear processes might be integrated out, the validity of this assumption needs to be tested.

perturbation strength The steady state assumption itself proves to be another critical point which is further complicated by the requirement of infinitesimally small perturbations (Kholodenko et al., 2002). In biological systems (quasi) steady states can only be approximated and perturbations need to surpass the level of noise to be experimentally detectable, rendering small perturbations impractical.

Therefore strong perturbations and imprecise steady state definitions are possibly reducing the predictive power of MRA. There are hints that larger perturbation strengths can be handled by MRA as it has been shown by theoretical modelling that perturbations of 10 and 50% result in the same predicted network in a three-tired model (Kholodenko et al., 2002). However, knockdown efficiencies achieved by RNA interference regularly surpass 50%, calling for a test of MRA on stronger perturbations.

incomplete data Classical MRA requires all nodes to be perturbed once at a time with subsequent measurement of all nodes. This is throughout feasible for almost all genetic networks by knockdown- or overexpression experiments with subsequent microarray or qPCR detection. However, in signalling experiments this is limited by the availability of detection antibodies and by the ability to interfere within the signalling cascade. Although in some cases modularisation might be an appropriate solution, in other cases this might lead to oversimplification or ambiguous modularisation possibilities (e.g in branching scenarios).

multiple perturbations A last aspect that requires further attention is to investigate whether the methodology can be further developed to also incorporate multiple simultaneous perturbations. As multi-drug therapy promises to be the next step in targeted cancer treatment, a routine that can predict outcomes *in silico* might greatly help to filter out the most promising combinations for further experimental investigations.

1.5 Objectives

In the last decades numerous promising modelling approaches have been developed that are suitable for modelling biological data. However, they differ in input requirements, underlying methodology and obtainable results. The aim of this thesis is - instead of inventing yet another reverse engineering technique from scratch - to rather choose a promising technique from the plethora of existing models and to extensively benchmark its abilities to assist in molecular systems biology. For this thesis the modelling basis was chosen to be Modular Response Analysis. This is due to its ability to handle important biological structures such as feedbacks and crosstalks as well as weighting connectivity strengths in a non-discrete manner by still requiring only manageable amounts of data with strong flexibility to the molecular accuracy levels (by simulating modules).

To be able to apply MRA to noisy experimental data sets, a maximum likelihood routine in combination with an iterative model selection strategy is added in chapter 2 that allows to distinguish between real and chance connections in $\mathbf{r}$. The extended MRA is then extensively assessed on *in silico* genetic networks for the behaviour to the change of basic network properties including perturbation strength, noise and nonlinearity and is afterwards compared to alternative approaches. As a practical example, a use case for the reconstruction of a genetic network in ovarian cancer research will be discussed.

In chapter 3 the extended MRA will be further refined to enable a parametrisation from incomplete data and to allow the simulation of multiple perturbations. This will be utilised to model the signalling network around the epidermal growth factor receptor (EGFR) in a colon cancer cell line panel and to predict the outcome of combinatorial perturbations.

To reverse engineer networks from biological networks adjustments of MRA are needed. In hypothesis testing this is not the case. Therefore in chapter 4 the potential of plain MRA in conceptual modelling is tested for a case where feedbacks are involved and parametrisation is too scarcely known to efficiently model with ODEs. For this, the study will analyse a conflicting observation of the effect of an RNA binding protein on the normoxic regulation of hypoxia induced factor 1 (HIF-1).

Reverse engineering of genetic networks

Parts of this work have been published in Parikh et al. (2010); Stelniec-Klotz et al. (2012) and Thomas et al. (2015). Contributions of co-authors are indicated.

Synopsis

Genetic networks represent the regulatory layer that control the cellular phenotype via changes of the RNA and protein composition. In this chapter I investigate the potential of MRA in predicting genetic networks. A maximum likelihood (ML) directed model selection (MS) procedure is connected to MRA (termed ML MS MRA) to account for noisy data. This methodology is benchmarked on artificial test sets mimicking various aspects of gene regulatory networks. Furthermore ML MS MRA is compared to other variants of MRA and a competing method and is shown to exceed or match their performances for many settings. The test sets showed that successful application of ML MS MRA is restricted to small sparse networks and to data with a distinct signal-to-noise ratio. Despite those limitations experimental use is demonstrated by illustrating a successful approach to reconstruct a signalling-driven genetic network downstream of oncogene activation.

2.1 Introduction

Signal transduction is commonly perceived as the means of the cell to relay external cues to the nucleus where its information is decoded into changes of gene expression. Although signalling pathways are comparatively conserved, the observable response strongly depends on the cellular context. Transforming growth factor-β (TGF-β) signalling for example can mediate mutually exclusive processes such as growth inhibition/proliferation or maintenance of stem cell pluripotency/differentiation (Massagué, 2012). Those heterogeneous responses are reflected by changes in gene expression patterns. Therefore for the eventual understanding of the molecular cell,

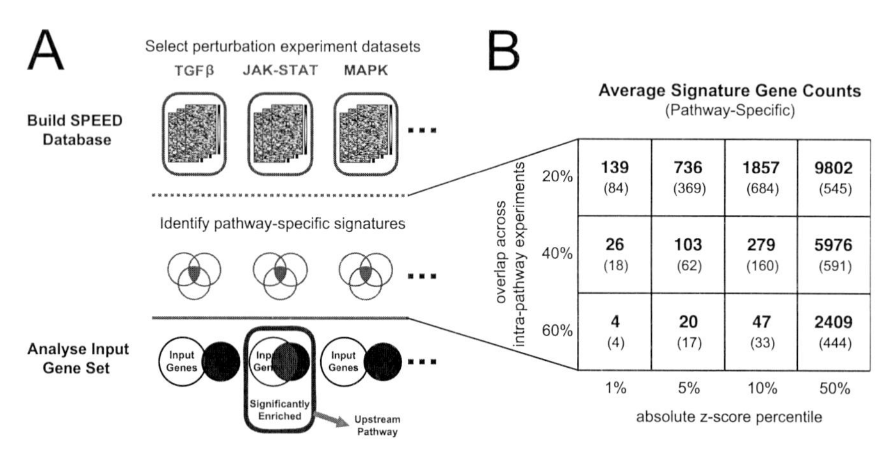

Figure 2.1 Pathway-specific signatures regularly contain shared genes (A) Workflow of the SPEED tool (Signaling Pathway Enrichment using Experimental Datasets). Collection of early microarray perturbation experiments (up to 4h) specific for 11 pathways. Generation of single-pathway signatures by assessing the intra-pathway overlap of differentially expressed genes. Users can now compare their dataset to the signatures and derive the causal upstream pathway(s). (**B**) Investigation of pathway-specific signature gene composition. Average number of signature genes for four well distinguishable pathways : JAK-STAT (14 exp.) , TGF-β (38 exp.), MAPK + PI3K (39 exp.) and TLR (39 exp.) for different cutoffs in z-score and overlap across intra-pathway experiments. Brackets denote average number of signature genes not shared with other pathways. *Taken from Parikh et al. (2012), Figures adapted of Figure 1 in (A) and combination of Figure 2 and Suppl. Figure 1 in (B), originals created by Jignesh Parikh*

it is crucial to be able to connect gene expression patterns to cellular responses both upstream and downstream.

A commonly applied method for downstream analysis is called Gene Set Enrichment Analysis (GSEA) (Subramanian et al., 2005). GSEA searches for overrepresented subgroups of genes by evaluating classification systems such as the gene ontology terminology that groups genes according to biological function, pathway membership and localisation (Ashburner et al., 2000). Although this approach has been proven useful in many applications to guide further research, some limitations exist. It is not distinguished between positive or negative influence of the gene, searches in categories with varying member counts is statistically challenging and the accuracy of the categories is highly variable as it mirrors the focus of the research community. In addition, the observed change in gene expression might not translate to the actual protein activity for which posttranslational modifications such as phosphorylations are often required. This illustrates that the prediction of cellular responses from gene expression patterns is rather complicated.

However, the reverse scenario, i.e. connecting causative signalling to the gene expression pattern, seems to be comparatively straight forward. In order to test this, we searched the GEO database for pathway or stimuli-specific responses ranging from 20 minutes to four hours. We found 215 experiments from which we were able to extract genetic footprints for 11 distinct pathways (Parikh et al., 2010). These pathway-specific signatures turned out to contain many genes that are shared between several pathways (Fig. 2.1B), however their individual composition is unique. For four pathways, where sufficient alternative literature-derived gene lists were available, we could demonstrate that the signatures are sufficient to reliably predict the causative upstream signalling. Most notably this included also the versatile TGF-β pathway. The resulting database called 'Signaling Pathway Enrichment using Experimental Datasets' (SPEED)[1] could present a suitable starting point when trying to disentangle microarray expression of healthy and cancerous tissue by unravelling altered upstream pathways (see Fig. 2.1A). The ambiguity of the pathway-assigned genes is demonstrated by noting that on average only four genes were left when demanding an overlap of 60% and a z-score within the top percentile for four distinct pathways (Fig. 2.1B). Moreover, when attempting to find gene sets that are exclusively regulated by only one pathway the number is further reduced for all but the most extreme threshold. When considering a wider range of the ≈ 30 classical mammalian signalling pathways listed in Kyoto Encyclopedia of Genes and Genomes (KEGG) (Kanehisa and Goto, 2000) it is apparent that for reasonable cutoffs the signature will contain almost no genes that are exclusively regulated by a single pathway. This supports the conception of signalling and genetic regulatory structure as an interlacing network rather than linear paths.

On the gene regulatory level the network propagation is commonly thought to be conducted by a transcription factor network. It has become clear that the transcription factor network does not represent a strict hierarchical network, but that it is highly interspersed with feedback and feedforward structures (Amit et al., 2007; Odom et al., 2004; Harris and Levine, 2005). The increasing availability of Chromatin immunoprecipitation (ChIP) experiments strongly enhanced by the ENCyclopedia Of DNA Elements (ENCODE) project (Consortium, 2004), has improved the definition of binding regions for many transcription factors. It has, however, emerged that this binding pattern depends on the cellular context and even on the signalling state of the cells (van der Meer et al., 2010). Therefore it

[1] http://speed.sys-bio.net/

is crucial to understand the particular genetic network acting under the present cellular circumstances.

To derive these genetic snapshots modular response analysis is thought to assist for two reasons: First, the currently most wide-spread genetic perturbation techniques, siRNA and RNA overexpression are fulfilling the requirement of steady state measurements as they are typically measured about 24-48 h after transfection[2]. And second, MRA is able to resolve the expected frequent occurrences of feedback and feedforward structures.

2.2 Statistical extension of MRA

To enable the application of the promising approach of MRA to real experimental data sets, adjustments are required that enable MRA to reliably distinguish between real regulatory interactions and noise. The easiest solution would be to set an arbitrary threshold for the r^i_j's to distinguish between noise and regulation, as has been done in Camacho et al. (2007) for the application of a variant of MRA called Regulatory Strength Analysis (RSA). However, a general threshold will fail to encompass the varying specificity of the detection molecules, e.g. probe specificity. Furthermore, linear noise thresholding can not account for the nonlinear propagation of noise terms that occur via inversion of $\mathbf{R}$. Therefore an alternative procedure had to be developed to distinguish between real and chance connections in $\mathbf{r}$. Since in a similar manner noise also hampers the direct derivation of the perturbation strength from the global response matrix, the procedure will have to encompass the estimation of $\mathbf{p}$ as well.

To tackle this problem, Nils Blüthgen and I developed an algorithm that estimates local response coefficients using a maximum likelihood estimation in combination with a model selection approach following a strategy described by Timmer et al. (2004). The approach, which is delineated in the following two sections will then be benchmarked to investigate its potential.

2.2.1 Maximum likelihood extension

By regarding the theoretical global response matrix $\mathbf{R_{model}}$ as a function of $\mathbf{r}$ and the perturbation diag($\mathbf{p}$), the main Equation (1.4) can be reformulated as:

$$\mathbf{R_{model}}(\mathbf{r}, p) = -\mathbf{r}^{-1} \cdot \mathrm{diag}(\mathbf{p})\,. \tag{2.1}$$

[2]This applies also to the next generation of genetic modulators such as the CRISPR/Cas system.

The goal was now to minimise the difference between the theoretical and the experimentally derived global response matrix $\mathbf{R_{data}}$. Therefore an objective function was developed that gives for each $\mathbf{r}^{-1}$ and diag($\mathbf{p}$) the negative logarithm of the likelihood:

$$\begin{aligned}\chi^2(\mathbf{r}^{-1}, \text{diag}(\mathbf{p})|\mathbf{R_{data}}) &= \sum_{ij} \left(\frac{\text{R}_{\mathbf{data}}{}^{i}_{j} - \text{R}_{\mathbf{model}}(\mathbf{r}, p)^{i}_{j}}{\epsilon^{i}_{j}} \right)^2 \\ &= \sum_{ij} \left(\frac{\text{R}_{\mathbf{data}}{}^{i}_{j} - (-r^{-1} \cdot \text{diag}(\mathbf{p}))^{i}_{j}}{\epsilon^{i}_{j}} \right)^2 . \end{aligned} \tag{2.2}$$

Here, ϵ^i_j is the error of the corresponding global response coefficient $\text{R}_{\mathbf{data}}{}^{i}_{j}$ that can be estimated from replicate experiments by an appropriate error propagation model (see Suppl. Section A 6 on page 125). The perturbation strength p_i can be estimated by setting the derivative of (2.2) to zero.

$$0 = \left| \sum_{j=1} \left(\frac{\text{R}_{\mathbf{data}}{}^{i}_{j} - (-r^{-1})^{i}_{j} p_i}{\epsilon^{i}_{j}} \right)^2 \right|'_{p_i} = \sum_{j=1} \frac{-2(-r^{-1})^{i}_{j}(\text{R}_{\mathbf{data}}{}^{i}_{j} - (-r^{-1})^{i}_{j} p_i)}{\epsilon^{i\,2}_{j}} . \tag{2.3}$$

Solving the thusly found global minimum for p_i results in:

$$p_i = \left(\sum_{j=1} \frac{(-\mathbf{r}^{-1})^{i}_{j} \text{R}_{\mathbf{data}}{}^{i}_{j}}{\epsilon^{i\,2}_{j}} \right) / \left(\sum_{j=1} \frac{(-\mathbf{r}^{-1})^{i\,2}_{j}}{\epsilon^{i\,2}_{j}} \right) . \tag{2.4}$$

Thus for any given $\mathbf{r}$, the corresponding perturbation strength vector $\mathbf{p}$ can be calculated by weighted least squares regression.

To find parameter sets that match the observed global response matrix best in a maximum likelihood sense, $\mathbf{r}$ (and indirectly $\mathbf{p}$) have to be selected such that they minimise the weighted sum of squared deviations of Eq. 2.2, which can be interpreted as χ^2-values. In practice, $\mathbf{r}$ can only be estimated numerically, by e.g utilising the Levenberg-Marquard algorithm (implementation used from Lourakis (Jul. 2004)). For numerical stability the inverse was approximated by the pseudoinverse using singular value decomposition.

2.2.2 Greedy hill-climbing model selection

Since it is of interest to determine the optimal number of non-zero entries of $\mathbf{r}$, one would ideally have to test all possible combinations which is computationally feasible only for very small networks. To circumvent the curse of dimensionality

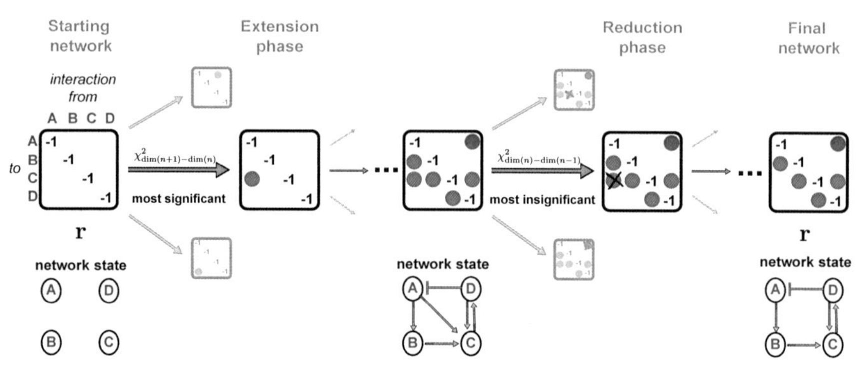

Figure 2.2 Model selection by ML MS MRA The procedure starts with a local response matrix **r** initialised as a negative unity matrix. During the extension phase every possible link is added separately and parameters are estimated by a maximum likelihood approach. For the link that improves the likelihood most, it is tested whether this improvement is significant or just due to increased model complexity. If significant, the link is kept and the prediction will be taken as the new standard to compare with when trying to add another link. This procedure is continued until no significant improvement can be found by any added link. Next, the networks are cleared of redundant links. In the so-called reduction phase one link at a time is removed to test whether the prediction would be significantly decreased. After no further redundancies are found the network prediction is completed.

a greedy hill-climbing model selection strategy was devised that is illustrated in Fig. 2.2.

Sparsity is enforced by starting with a local response matrix as a negative unity matrix ($\mathbf{r} = -\mathbf{I}$), i.e. a completely unconnected network. Subsequently it is tested systematically for each possible link whether addition significantly improves the model fit, as judged by a likelihood ratio test (Timmer et al., 2004). More precisely, the difference between the χ^2-value of the unaltered model and the χ^2-value of the best-fitting altered model is compared with a χ^2-distribution with one degree of freedom. If the difference in χ^2-values of the nested models translates into p-values below alpha ($\alpha = 0.05$), adding the link will significantly improve the fit. This procedure is repeated until no further link significantly improves the fit. In the course of the extension procedure sometimes redundant paths are added which would not significantly improve the fit when added in a different order. Thus after completion of the extension phase the network is iteratively pruned off redundant links. During this phase, links are removed if the likelihood is not significantly reduced. This reduction phase is carried out until no superfluous links are contained in **r**.

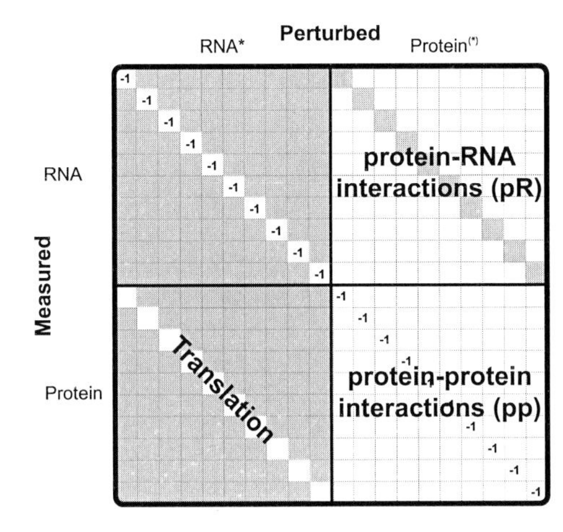

Figure 2.3 Test set structure in local response matrix format After the proteins are translated from their corresponding RNA they either act as transcription factors (protein-RNA interaction), or regulate the half-life of other protein species (protein-protein interaction). For each column the ith entry is perturbed (RNA directly and protein indirectly) and the change in steady state is measured for all species. Grey areas denote forbidden interactions prohibiting RNA interactions other than translation and regulation of RNA by its encoding protein.

Because the method combines modular response analysis (MRA), maximum likelihood (ML) parameter estimation and model selection (MS), the resulting algorithm is termed ML MS MRA. Within this framework, introduction of previous knowledge is straight forward: If links are known, they can be introduced in the starting network, and if links are biologically not reasonable, the corresponding entry in $\mathbf{r}$ can be fixed to zero.

2.3 Parameter benchmark on *in silico* networks

In order to investigate the predictive capability and robustness of the extended MRA algorithm, artificial genetic networks were generated and evaluated. Moreover, the performance was compared to other established reverse engineering approaches used in the literature.

2.3.1 Artificial gene regulatory network generation

The test sets were chosen such that they most closely resemble real experiments in structure, dynamics and parameter distributions. The default networks consist of 10 species of mRNA and their corresponding proteins which can interact as depicted in Fig. 2.3 in the typical MRA perspective, i.e. perturbed nodes in the columns and measured nodes in the rows. Proteins are allowed to act in two ways (i) as transcription factors or (ii) as posttranslational modifiers of proteins depicted on the right upper and lower quadrant, respectively. For simplicity it is assumed in here that RNA serves only as template for translation and exhibits no ribozyme activity (indicated by greyed out regions in the left half). Systematic perturbations were only acting on mRNA level, mimicking classical siRNA or overexpression experiments. Thus in the system not all nodes can be directly perturbed but all signals can be measured. Together with the assumption that RNA can only translate to the acting protein, perturbations on the RNA level are expected to correlate with changes of the protein level in steady state.

The underlying dynamic model was generated using an ordinary differential equation system based on work of Mendes et al. (2003). Two types of regulations effecting the concentration were included: transcriptional regulation of mRNAs and posttranslational modification of proteins - all regulated by proteins inside the network. The concentrations of mRNAs R_i and proteins P_i were modelled such:

$$\begin{aligned} \frac{dR_i}{dt} &= V_{\max,R_i} \cdot \prod_j \left(\frac{P_j^{h_j}}{P_j^{h_j} + K_j^{h_j}} \right) \cdot \prod_k \left(\frac{K_k^{h_k}}{P_k^{h_k} + K_k^{h_k}} \right) - d_{R_i} \cdot R_i \\ \frac{dP_i}{dt} &= R_i \cdot V_{\max,P_i} - d_{P_i} \cdot P_i \cdot \prod_j \left(\frac{K_j^{h_j}}{P_j^{h_j} + K_j^{h_j}} \right) \cdot \prod_k \left(\frac{P_k^{h_k}}{P_k^{h_k} + K_k^{h_k}} \right) . \end{aligned} \tag{2.5}$$

$V_{\max,R_i}$ is the basal transcription rate, which is changed by knockdown or overexpression experiments, i.e. the parameter altered to simulate perturbations. For RNA production the transcription rate is increased or decreased by activatory (first product) and inhibitory (second product) actions of the network proteins j, k that work as transcription factors. Similarly, the protein degradation rate d_{P_i} can be altered in the production term, mimicking protein-protein interactions. K represents a constant in a similar fashion as the Km value in a Michaelis-Menten equation (Mendes et al., 2003) and h is the Hill coefficient representing the factor of cooperativity. In this model it is assumed that no protein can inhibit and activate the same RNA species ($j \neq k$). Furthermore, self regulations of either RNA or protein levels were excluded and therefore $i \neq j, k$.

To generate MRA compatible data sets steady states had to be derived. First the steady state of the unperturbed system was approximated after 1000 iterations of Eq. 2.5 with starting parameters set to 1. Then, $V_{\max,R_i}$ was changed and the network was allowed to iterate 1000 times to reach the post-perturbation steady state. That this setting will not always allow to reach the actual steady state is part of the applicability challenge as also biological systems are rarely in absolute steady state. After steady state approximation multiplicative Gaussian noise, representing the biological error, was added to both data sets (default 20%). Based on experimental implications test set parameters were drawn from log-normal distributions (for a detailed description of parameter quantifications see Suppl. Section A 1 on page 113).

2.3.2 Parameter benchmark of refined MRA approach

The artificial networks are now subjected to be reverse engineered from scratch by ML MS MRA. To investigate the optimal experimental settings, parameters were varied for external (perturbation strength, network size) as well as internal (noise, non-linearity, sparseness) characteristics. The respective default network consists of a 10 node network with 20% link density and mild nonlinearity ($h = 2.5 \pm 0.7$) with perturbation strength of -50% and noise set to 20%. These default settings are marked by grey arrows in the corresponding figures.

The predictions were evaluated by interpreting three statistical measures: (i) Matthew's correlation coefficient (mcc), (ii) sensitivity, and (iii) precision (mathematical definition in Suppl. Chapter A 2). Matthew's correlation coefficient reflects the overall performance and was chosen due to its reported superiority among single-number classifiers (Baldi et al., 2000). In addition the interpretation of mcc is straight forward. In analogy to correlation coefficients 1 indicates perfect reproduction, 0 no better than random, and -1 stands for absolute disagreement. Thus by mcc it can be directly assessed whether the prediction performs better than random guessing. Sensitivity describes the depth of link recovery and precision defines the validity of all predicted links. These specific aspects are chosen since they are of importance to assess the quality and quantity of the prediction.

One of the major problems when trying to predict gene regulatory networks is that the changes seen on RNA level might not propagate linearly to the protein level. In mice cell lines it has been observed that mRNA can explain at most 40% of the differential expression of their respective proteins (Tian et al., 2004; Schwanhäusser et al., 2011). Also a study of the human cancer cell line panel (NCI-60) reported a Pearson correlation coefficient of 0.42 (Shankavaram et al., 2007), which further was stated to be different for proteins with structural function (0.71) and those involved in signal transduction (0.39). This is similar for the steady state correlation in the artificial test sets (cf. Suppl. Fig. A.2). Therefore it seemed appropriate to test the applicability of MRA on measurements encompassing both mRNA level and protein level to provide a more optimal study environment for the parameter evaluation.

With transcriptional and posttranslational regulation, two qualitatively different network interactions are included. The latter interaction type was chosen to monitor MRA capability to predict indirect interactions, as perturbations were set to only act on RNA level. To determine differences in predictivity, networks containing only one regulation type were tested in addition to the combined scenario. The results for those three cases are exemplarily shown in Fig. 2.4. Due to the similar trend in the three network types, afterwards only transcription factor networks (TF) as the most important subgroup will be depicted. However, in Suppl. Section A 3 on page 117 boxplots for all scenarios are provided for all parameter benchmarks as well as performance on sole RNA data.

perturbation strength MRA is defined to be most accurate if perturbations are small enough to remain in the original basin of attraction, i.e. smaller perturbations should yield better predictions. However, if the perturbations are small, the magnitude of the response may be smaller than the noise fluctuations. Therefore, a trade-off between perturbation strength and noise is to be expected. To investigate this systematically, *in silico* data sets with different strengths of perturbation were generated, i.e. changing V_{max,R_i}, while leaving noise constant.

Fig. 2.4 shows that there is a general tendency that stronger perturbations increase performance of the algorithm for all three statistics. Interestingly, knockout-like conditions[3], allow the best predictions, whereas predictions from 10% knockdown data give unreliable results. This inability of prediction can most likely be attributed to the low signal-to-noise ratio (average noise was set to 20%). It can further be noted that a two-fold perturbation in either direction (-50% and +100%) yields about the same results. Therefore, not perturbation direction, i.e. knockdown or overexpression, but perturbation strength determines the performance. To conclude, although MRA relies on linearisation, strong perturbations enable the algorithm to detect more interactions with a higher precision.

The observed tendencies are conserved in all three network types, although it can be noticed that in comparison to the single interaction type scenarios, the overall performance of the mixed case is lower. This might be due to the increased parameter space as in the mixed model the same number of interactions is distributed to both transcriptional (pR) as well as posttranslational (pp) regulation (see also Fig. 2.7). In addition, when comparing the predictions of networks containing only transcriptional interactions (pR) versus networks with posttranslational interac-

[3]Due to the model design a direct KO was not feasible, instead it was approximated by multiplying V_{max,R_i} by 10^{-6}

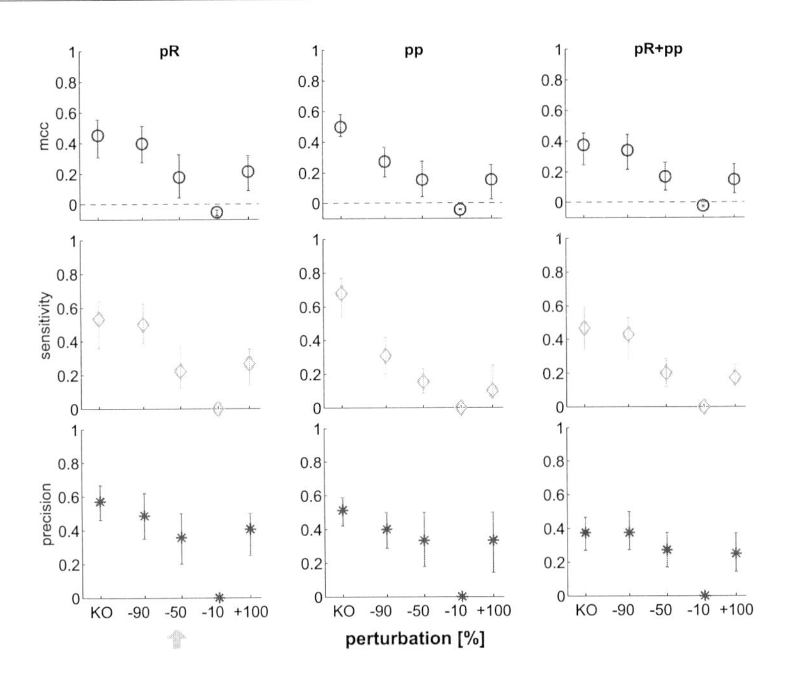

Figure 2.4 Absolute perturbation strength but not direction effects ML MS MRA performance Evaluation of ML MS MRA prediction on 100 simulated ten node networks for different perturbation strengths. The initial perturbation, indicated as percent change of the transcription rate, includes an approximated knockout (KO) and overexpression (+100) scheme as well as three knockdown scenarios. Method performance is depicted as medians with upper and lower quartiles for Matthew's correlation coefficient (mcc), sensitivity and precision. Three network scenarios with the same number of interactions (18) were investigated: **left** protein-RNA (pR), **middle** protein-protein (pp) and **right** a combination of both interaction types (pR+pp). The arrow points to the default parameter value that is used when varying other settings in the following figures.

tions (pp) performance seems slightly worse for all cases but KO where sensitivity is even higher. Therefore ML MS MRA is capable to infer indirect interactions for additionally measured but not perturbed nodes almost as well as for nodes that are directly perturbed.

noise From the previous paragraph it can be expected that noise in experimental perturbation studies will play a critical role, with more noise leading to worse predictions. Indeed, when changing the noise level and keeping perturbation constant at -50% it can clearly be seen from Fig. 2.5A that the overall performance is declining. This decrease in mcc is due to the decreased median sensitivity (from

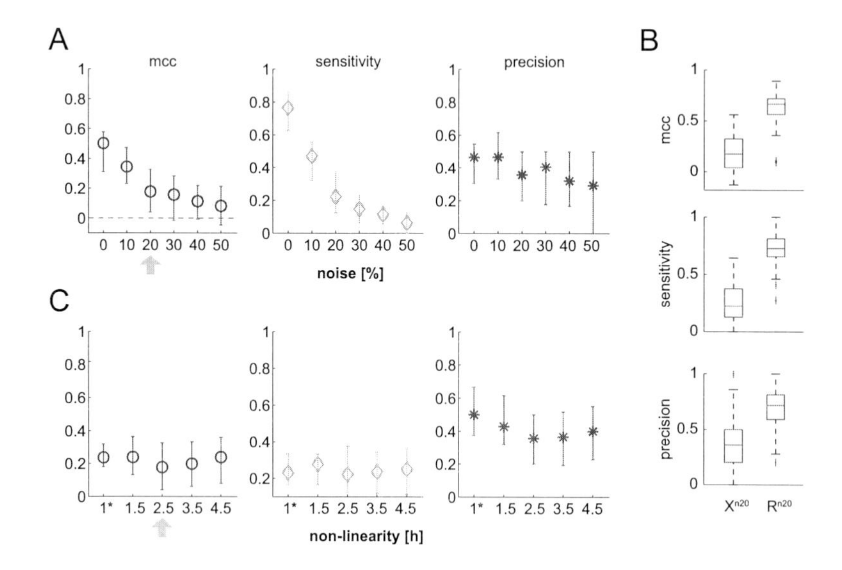

Figure 2.5 Noise but not cooperativity limit performance (**A**) Median statistical performance in response to different noise levels exemplified on transcription factor networks. (**B**) Boxplots comparing the predictivity with 20% noise added to the measurements (X^{n20}) or added directly to the global response coefficients derived from unnoised measurements (R^{n20}). (**C**) Median statistical performance for different hill coefficients as a proxy for varying non-linearity. The x-axis displays the average h with all $\sigma_h = 0.71$. The asterisks denote the linear case with $h = 1$. See also Suppl. Fig. A.4 on page 118.

77% for no noise to 6% for 50% noise), while the median precision is much less changed (from 47% to 29%). This indicates that the quality of the prediction is comparatively robust against noise which comes at the cost of less recovered links. In light of the insights drawn from the previous paragraph it can be stated that the performance is best when noise is minimised and the initial perturbation is maximised.

Of concern is the low precision hinting to a high percentage of wrongly predicted links which might be due to the noise propagation. To investigate this, the prediction for the same data set was compared, once with the measurements noised by 20% and once where the global response coefficients were calculated from the unnoised steady states and noise was applied afterwards (Fig. 2.5B). It can be noted that the predictions on the noised R (R^{n20}) show a better sensitivity and more importantly a much improved precision (from 36% to 72%). This is most interesting as sensitivity in here is comparable but precision is much higher than that of the prediction of the noise-free dataset[4] depicted in Fig. 2.5A. The reason must reside in the different assessments of the error. In the noise-free case the error term is very low and therefore the maximum likelihood approach will detect more links to significantly improve the fit. Among the large number of recovered links, however, many false positives seem to be present which is not the case when the error model is sufficiently conservative as in Fig. 2.5B. Therefore the maximum likelihood tends to overfit if the error term falls under a certain threshold, supporting the use of a more conservative error estimation for higher precision. The much better performance on the global response coefficient that has been created from unnoised steady states shows that the strongest confounder is not the noise level on $\mathbf{R}$ but the deviation of steady state estimations from the true steady state. A solution to increase reliability of steady state estimations is to repeat the measurements as the standard error of the mean is reduced by the square root of repeat measurements. For example from 4 repeated measurements with standard deviation of 20% the standard error of the estimated mean will be on average reduced by half (10%). Thus a way to improve MRA overall performance is to use replicate measurements thereby increasing the accuracy of the steady state assessments.

non-linearity The basis of MRA is a linearisation of the underlying ODE system around its (unique) steady state. If the system is non-linear, this approximation might not apply.

In order to investigate the effect of non-linearity, the cooperativity coefficient h from Eq. 2.5 was varied (Fig. 2.5C). Neither increasing h_μ to 4.5 nor assuming linear interactions (constant $h = 1$) could significantly affect the overall performance (mcc). It can, however, be noted that the prediction tends to be more accurate for less non-linear relationships (50% for $h = 1$ to 36% for $h_\mu = 2.5$) whereas sensitivity remains similar. Therefore it appears that under the tested parameter settings ML MS MRA is not strongly affected by non-linearity in the interaction terms. As a side effect the size of the apparent r_j^i's is increasing with increasing hill coefficient (cf. Suppl. Fig. A 1 on page 113B). However, as the model predictions do not increase, absolute strength of r seems to be of minor influence.

network size Recently, quality and automation of knockdowns as well as measuring techniques have evolved such that reverse engineering tasks for genetic networks can nowadays vary in size from a few nodes to hundreds or even thousands

[4]Due to the maximum-likelihood approach a noise level of 0 was not feasible, instead a factor of 10^{-6} was used

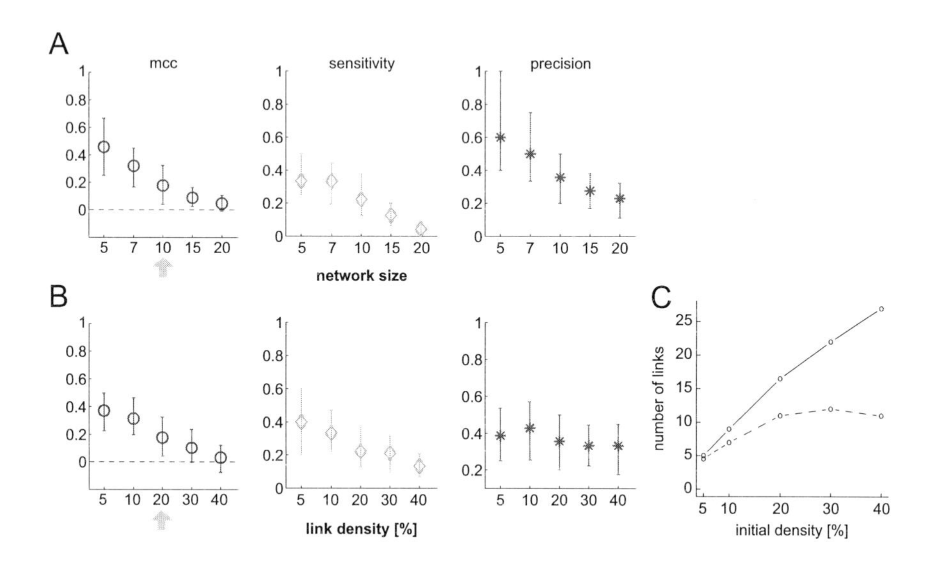

Figure 2.6 Small and sparse networks yield generally better predictions Median performance for (**A**) networks of different sizes and (**B**) networks exhibiting differing percentage of link densities. (**C**) Comparison of the median number of real non-zero links in the network (continuous) and the median number of predicted links (dashed) for the networks investigated in (B). See complete set in Suppl. Fig. A.5.

of nodes. Because of computational limitations, owing to the model selection procedure, only small scale networks could be investigated ranging from 5 to 20 node networks. Fig. 2.6A shows that the proposed method performs better on smaller networks for both quantity as well as quality. For networks as large as 20 nodes the sensitivity approaches zero (4%) with the precision dropping to 23%. In here a weakness of the local optimisation strategy can be spotted, which seems to be unable to find an acceptable solution for larger networks. Thus this methodology should only be considered for small scale networks and will not be useful to model genome-wide networks as long as the greedy hill-climbing approach is not improved or replaced by a global optimisation strategy.

sparsity Next to network size also the network density is expected to have an impact on the reverse engineering capacities of ML MS MRA. One would assume that a sparse network will generally perform better than a dense network as the number of feedback and feedforward structures is expected to be lower. Also, a less dense network will produce a smaller number of recovered links limiting the error possibility that accompanies the chosen local model selection strategy. Indeed in

Fig. 2.6B a higher sensitivity could be found for more sparse networks. However, precision remains largely invariant especially for networks with densities larger than 10%. This stands in contrast to the case when network size was varied where both sensitivity and precision was reduced. By comparing the real average number of non-zero links with the average number of predicted links it can be noted that the latter stagnates for higher link densities (Fig. 2.6C). This explains why a smaller fraction of links is found in more dense networks. On the whole, the hypothesis that sparse networks are better represented can be confirmed.

different data sources: RNA vs RNA + protein One improvement to conventional MRA is that ML MS MRA can be run on more complicated structures than square matrices allowing to provide additional information from measured but not perturbed nodes as well as to prohibit links by enforcing entries of $\mathbf{r}$ to stay zero. To investigate the impact of additional information, the performance of ML MS MRA on the same test sets was compared to the case when only predicting from

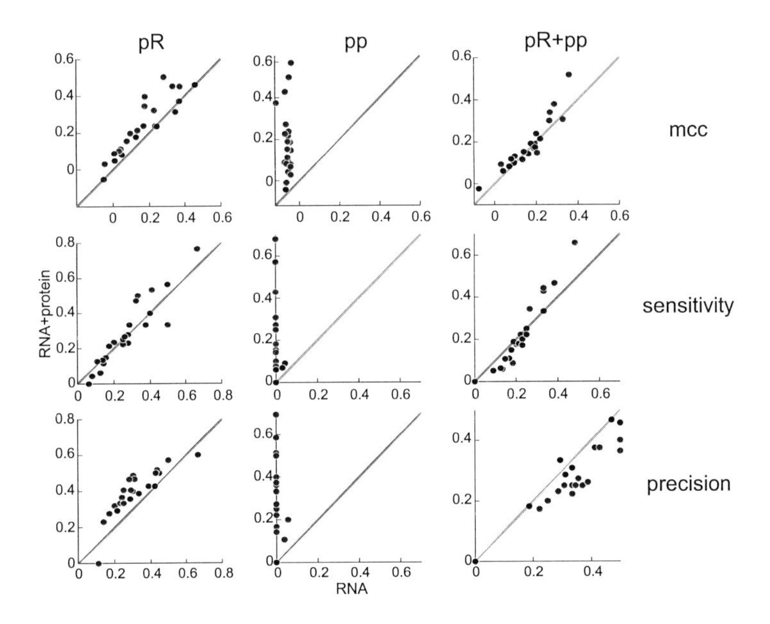

Figure 2.7 Additional information assertion Comparisons of median performance of ML MS MRA for all test sets generated for parameter benchmarking (listed in Suppl. Section A 3 on page 117) for three network types. Each scatterplot shows the deviation from parity (grey line) for predictions performed on RNA data (x-axis) or RNA + protein data (y-axis). Abbrev: pR=protein-RNA; pp=protein-protein interactions.

RNA data (Fig. 2.7). Both transcriptional as well as posttranslational networks predicted from RNA plus protein data exhibit regularly a higher mcc. In TF networks this is predominantly achieved by a better precision, with the exception of two cases: the unpredictive case $p = 10\%$ (lower left point) and the five node network case (upper right point) where presumably the low number of links may obstruct better prediction from more data. For the posttranslational network case no feedbacks on transcription are present and thus the prediction of interactions is only possible if protein data is available. This changes in the mixed scenario where even RNA is predictive for posttranslational modifications. While exhibiting comparable levels of the mcc, sensitivity is higher with additional information when overall sensitivity is high but precision is mostly worse than RNA-based predictions. The apparent drop in the performance for RNA+protein data in the mixed scenario is due to the fact that the underlying statistics is changed. If only RNA is given, the prediction can not distinguish between transcriptional or posttranslational regulation, e.g. a 10 node network without self regulation offers 90 possible links. If protein information is given these two interactions can be distinguished and by also requiring translation to be detected the link space is more than doubled to 190 possibilities. Therefore in order to obtain a benefit, the added information should not proportionally increase model complexity.

2.4 Comparison to alternative methodologies

After the general characterisation of ML MS MRA, the performance was compared to established reverse engineering methods. The alternatives had to bear the same characteristics as the introduced method, i.e. complying with steady state perturbation data and being able to predict cyclic networks. From the remaining candidate methods I decided to test and compare three approaches: NIR, plain MRA and an MRA variant connected to Monte-Carlo (MC) sampling which was consequently termed MC MRA. NIR and MC MRA were selected because they represent the only methods that have been successfully applied to experimental data sets. Plain MRA was included to investigate whether the extensions in ML MS MRA have affected the overall performance. Below a short introduction of the main model features are given. For a detailed explanation of the methodology and description of application please refer to Section A 4 in the Appendix.

2.4.1 Established steady state prediction methods

plain MRA follows the original derivation of **r** by Kholodenko et al. (2002) as formulated in Eq. 1.5. To distinguish between real and chance connection, a threshold for the absolute magnitude of the local response was set.

MC MRA extends this original concept by using stochastic sampling of the global response matrix by applying Gaussian noise to the original **R**. Through the sampling a confidence estimation for the local response coefficients can be utilised for thresholding, thus replacing the cutoff of plain MRA. This methodology was first introduced by Andrec et al. (2005) and extended and successfully applied on a signalling network around the MAPK-cascade in Santos et al. (2007).

NIR was first proposed by Gardner et al. (2003) and estimates the unscaled Jacobian matrix by an ordinary least squares approach. In contrast to MRA, in NIR it is assumed that due to noise even a fully perturbed system is underdetermined and thereby the information only suffices to derive a sparse network. This sparsity is enforced by only allowing a subset of maximal $k < number\ of\ nodes$ non-zero input edges for each node. The most probable subset is derived by calculating all possible subsets for all k's by multiple linear regression and selecting the combination (and k) that can best explain the data. In practice, NIR has been successfully applied to reverse engineer transcription factor networks in *E.coli* (Gardner et al., 2003) and in a yeast synthetic network (Cantone et al., 2009).

2.4.2 Performance on default settings with different interaction types

The described methods were now compared to the newly proposed ML MS MRA on the *in silico* networks. First, the investigation was focussed on different interaction types which were compared for default parameter settings (marked by grey arrows in the figures of Section 2.3.2). For better comparability and lack of consistent cutoff measures the evaluated network connections were limited to the number detected by ML MS MRA, i.e. all networks for the same test set contained the same number of recovered links which were selected after the methods' inherent scoring system (see Suppl. Section A 4 on page 120). Since plain MRA and NIR were incapable to incorporate measured but not perturbed data, all methods were initially tested on RNA only data (bars without outline in Fig. 2.8). Afterwards MC MRA as well as ML MS MRA that are capable to model unperturbed data were provided with RNA and protein data (bars with black outline).

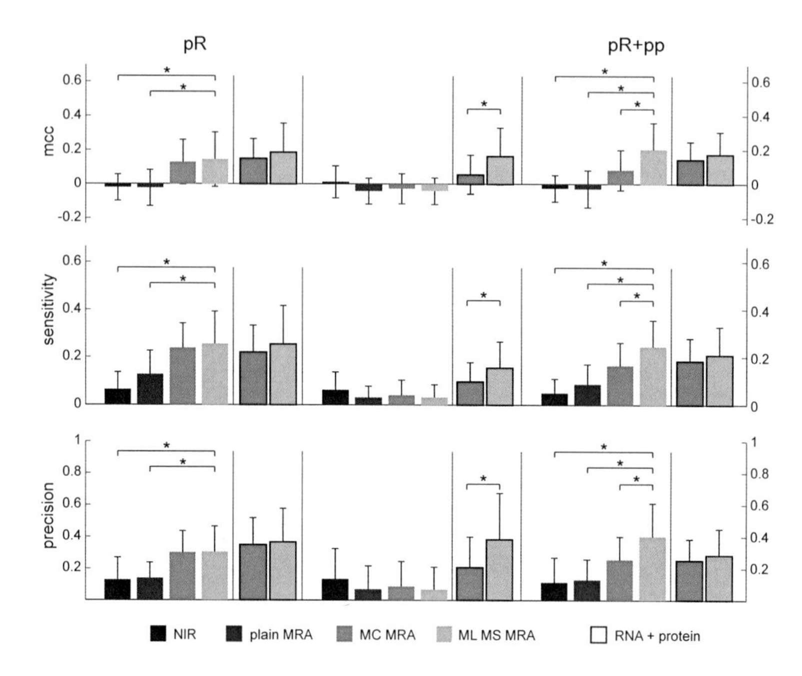

Figure 2.8 Comparison of steady state reverse engineering methods for different interaction types Average performance of NIR (black), plain MRA (dark brown), MC MRA (light brown), and ML MS MRA (yellow) on the *in silico* networks with default parameters (error bars = SD). Three different scenarios are depicted: **left** transcriptional regulation (pR) **middle** posttranslational regulation (pp), and **right** a mix of both (pp+pR). Predictions are based on RNA data (bars without outline) or RNA+protein data (bars with black outline). Asterisks denote significant differences of ML MS MRA estimated by a two-sided U-test ($p \leq 0.05$) with bracket ends indicating the compared entities.

In the middle column of Fig. 2.8 it can be asserted that when RNA is provided no method is capable to predict posttranslational regulation. Only when protein information is added a prediction better than random can be achieved (black outlined bars). This consolidates the findings from Fig.2.7 in the previous section.

When networks are predicted from RNA data, ML MS MRA significantly outcompetes plain MRA and NIR. Interestingly, MC MRA is significantly worse in the mixed scenario but performs similarly well on transcription factor networks. When RNA and protein data is provided, ML MS MRA prediction seems to be always slightly better than MC MRA, albeit only for posttranslational networks this advantage is significant.

Overall it can be stated that ML MS MRA seems to have improved MRA to be more versatile for different interaction types. From the found results it seems that ML MS MRA is the better choice when perturbation-independent interactions are present, i.e. posttranslational regulations. For transcriptional regulation, however, MC MRA performs equally well in the default test set. As this interaction type is the most basal in genetic network modelling a more closer inspection on parameter dependencies for those two methods was conducted.

2.4.3 Detailed method assessment on transcription factor networks

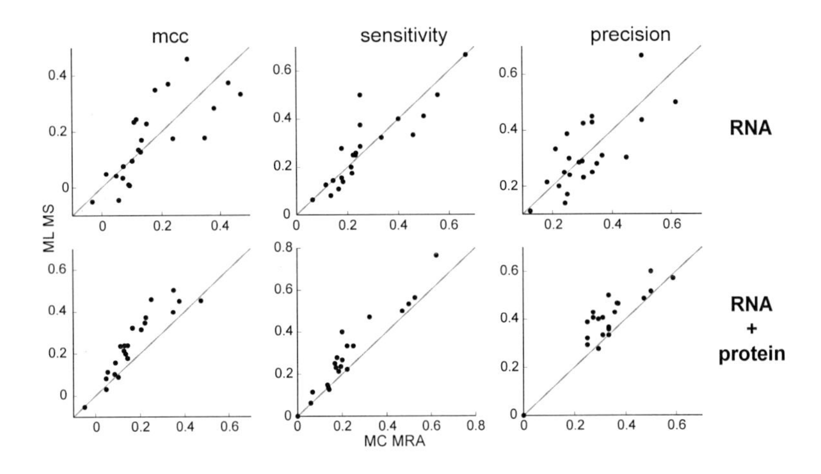

Figure 2.9 Median performance of MRA extensions Systematic comparison of median performance of ML MS MRA and MC MRA for all transcription factor test sets generated for benchmarking. Comparison on two different data sources indicated on the right.

In order to investigate in more depth which of the two extensions of MRA would produce the best predictions for transcription factor networks, I decided to compare the performance over all benchmarked conditions. The median performance for MC MRA and ML MS MRA was assessed on all transcription factor test sets generated for the parameter benchmarking (Fig. 2.9).

When predicting from mRNA-only data, a comparable number of cases shows better performance of either method. Especially for test set cases with higher median mcc scores the difference between prediction of the respective superior and inferior method seems to be large. When additional protein information is given this balanced scheme vanishes and it can be seen that ML MS MRA performs hardly worse and regularly surpasses MC MRA performance for all settings (cf. Fig. 2.9 bottom row). Thus, ML MS MRA seems to be the most fit prediction method, tested in this framework - also for transcription factor networks as long as additional information is given. When only RNA data is available both methods can be applied and the best method should be chosen according to the perceived circumstances.

In order to analyse these differences directly, the identity of the varied parameters was connected to the performance on RNA data in Tab. 2.1. From this some rules for the selection of the best method can be inferred. Regarding sparsity and network size ML MS MRA predictions excel (highlighted in yellow) when below the default settings whereas MC MRA performs slightly better (highlighted in blue) when above the default parametrisation. Since the slightly better performance for MC MRA for large and dense networks is only marginally better than random, one, however, might not recommend the usage of either MRA extension under the settings favouring MC MRA. Next to the seesaw relationship of network size and sparsity, for the remaining parameters only one method is favoured for some

Table 2.1 Comparing two extensions of MRA on whole RNA benchmark set Median performance of the two extensions of MRA: MLMS and MC on all *in silico* transcription factor test sets generated for parameter benchmark. Bright yellow highlighted fields denote better performance of ML MS MRA ($\geq$ 0.1)and bright blue of MC MRA. A fainter yellow and blue denotes smaller deviations larger than 0.03 for the same relationships.

parameter	value	mcc			sensitivity			precision		
		MLMS	MC	Diff.	MLMS	MC	Diff.	MLMS	MC	Diff.
default		0.13	0.13	0.00	0.25	0.22	0.03	0.29	0.29	0.00
density	5%	0.37	0.22	0.15	0.40	0.40	0.00	0.33	0.21	0.12
	10%	0.35	0.18	0.17	0.38	0.25	0.13	0.39	0.25	0.14
	30%	0.03	0.07	-0.04	0.17	0.22	-0.04	0.23	0.30	-0.07
	40%	-0.05	0.06	-0.10	0.14	0.18	-0.04	0.25	0.33	-0.08
nodes	5	0.46	0.29	0.17	0.50	0.25	0.25	0.67	0.50	0.17
	7	0.23	0.15	0.08	0.29	0.25	0.04	0.43	0.33	0.10
	15	0.01	0.09	-0.09	0.11	0.16	-0.06	0.17	0.25	-0.08
	20	0.01	0.09	-0.08	0.08	0.13	-0.05	0.14	0.24	-0.10
hill coeff.	1	0.24	0.12	0.13	0.28	0.18	0.10	0.45	0.33	0.12
	1.5	0.23	0.11	0.13	0.28	0.18	0.10	0.42	0.30	0.12
	3.5	0.10	0.10	-0.01	0.20	0.21	-0.01	0.24	0.26	-0.02
	4.5	0.17	0.13	0.04	0.25	0.23	0.02	0.30	0.26	0.04
perturbation	KO	0.33	0.47	-0.14	0.41	0.50	-0.09	0.50	0.62	-0.12
	-90%	0.18	0.35	-0.17	0.33	0.46	-0.12	0.30	0.45	-0.14
	-10%	-0.05	-0.03	-0.02	0.06	0.06	0.00	0.11	0.13	-0.01
	+100%	0.14	0.12	0.01	0.26	0.23	0.03	0.29	0.30	-0.01
noise	10^-6	0.28	0.38	-0.10	0.67	0.67	0.00	0.28	0.35	-0.07
	10%	0.18	0.24	-0.06	0.32	0.33	-0.01	0.31	0.37	-0.06
	30%	0.08	0.07	0.00	0.15	0.18	-0.02	0.25	0.24	0.01
	40%	0.04	0.05	-0.01	0.14	0.14	0.00	0.20	0.22	-0.02
	50%	0.05	0.02	0.03	0.13	0.11	0.01	0.21	0.18	0.03

settings. Whereas ML MS MRA produces more reliable networks for smaller hill coefficients, MC MRA performs better for strong perturbations and the 'unnoised' case. Thus both extensions complement each other and when applied corresponding to the found rules can produce more valuable results.

2.4.4 External benchmark: DREAM 4 network challenge

As the artificial networks tested and evaluated in here present only one of many attempts to simulate and assess genetic regulatory networks, a bias towards ML MS MRA might exist. Therefore another analysis on externally created genetic networks with a different statistical evaluation scheme was conducted.

The Dialogue for Reverse Engineering Assessments and Methods (DREAM) initiative, round 4, challenge 2 (Marbach et al., 2009; Stolovitzky et al., 2009, 2007) represents a valid framework to test predictions of transcription factor networks from RNA data. The challenge encompasses the prediction of five ten-node networks from five different data sets: steady state data for (i) wild type, (ii) knockdown (-50%), (iii) knockout (-100%) and (iv) multifactorial perturbations (all genes perturbed) and (v) five time series of unspecified perturbations affecting a third of the nodes. The networks were modelled by stochastic differential equations (SDE) with structure and parameters selected from existing networks in *Escherichia coli* and *Saccharomyces cerevisiae*. As part of the challenge all networks were selected to incorporate feedback structures. The task was to predict the directed unsigned topology by providing a ranking for all 90 possible interactions. This ranking was used to create Receiver Operation Curves (ROC) and Precision-Recall (PR) curves which were scored by weighing the respective area under the curves[5].

Table 2.2 depicts the results for ML MS MRA, MC MRA and plain MRA without threshold (MC MRA by confidence, ML MS MRA by order of link addition and plain MRA by absolute strength of r). When predicting from the knockout data set all methods would be ranked as third among the participants with MC MRA achieving a higher score than plain MRA and ML MS MRA. The high ranking position in the DREAM challenge suggests that MRA-based predictions from steady states represent clearly a valid prediction method. Even more so when acknowledging that the winner of this challenge used a Petri-net approach that predicted from all five available data sets (Küffner et al., 2010), which contains arguably more information than predictions from a single data source.

[5]Details on scoring can be found at https://www.synapse.org/#!Synapse:syn3049712/wiki/74628

Method	Score	Rank
Team 543	**7.13**	**1**
Team 549	**5.29**	**2**
MC KO	4.79	3
Plain KO	4.28	3
MLMS KO	4.17	3
Team 532	**3.97**	**3**
...	...	...
Team 161	**2.278**	**19**
MC KD	2.16	20
MLMS KD	1.24	25
Plain KD	0.85	27
Team 540	**0.5**	**29**

Table 2.2 DREAM 4 challenge 2 results for MRA variants Overall score and ranking among the 29 DREAM challenge participants (bold entries) of the three MRA methods ML MS MRA, MC MRA, and plain MRA when applied on knockout data (KO) or knockdown data (KD). The score generated by the DREAM evaluation software is based on the area under the receiver operation (AUROC) and area under the precision recall (AUPR) curves (cf. Suppl. Fig. A.6) for the prediction of five networks. Note that the challenge provided KO, KD, time series and multifactorial perturbation data, of which only the first two could be used by MRA.

The external challenge reproduces previous findings that MC MRA provides better results for strong perturbations. However, also the knockdown experiment which produced comparable results in the previous benchmark showed a better prediction score for MC MRA (but not plain MRA). The reason could reside in the fact that the error of the estimation of the pre-perturbed steady state $x_{(0)}$ was reduced as it was calculated from six replicate measurements (five zero time points of the time series data combined with wild type data set). This might have reduced the noise effect from which MC MRA seems to have profited the most. In agreement with previous findings small perturbations produced worse predictions than large perturbations. Interestingly combining both data sets did at best reproduce the knockout performance (not shown). Therefore combining data of very different quality does not improve predictions.

As the DREAM challenge score weighted the complete ranking, the actual quality of the produced networks are confounded by the insignificant links. Therefore the statistics were computed for the method's inherent cutoff (see Suppl. Sec. A 5). From this it was evident that MC MRA did not always perform best. Instead MC MRA performed only strongly superior in two of the five networks, whereas for one network predictions were equal and in two networks ML MS MRA performed better. Thus the previously found dependence on network characteristics has to be considered in order to choose between MC MRA and ML MS MRA. The original MRA performed equally well for the knockout data set and can be considered an applicable alternative if the cutoff may be differently chosen than from the strength of r, e.g. by considering the number of recovered links by ML MS MRA (cf. Suppl. Sec. A 5). In general the results from the two in silico challenges agree in many aspects and also the average performance of MC MRA is in a similar range while ML MS MRA shows higher sensitivity for the DREAM challenge approach (cf. KO data in Suppl. Tab. A.1 on page 125 and Tab. 2.1 on page 34).

In conclusion, without knowledge of the network settings, it cannot be stated beforehand which MRA approach will deliver the most reliable results from mRNA-based perturbation data.

2.5 Discussion

Modular response analysis represents a potentially powerful framework to recover an underlying network structure from systematic perturbation data of medium-sized networks. However, some obstacles had to be overcome to efficiently apply this concept to a wider range of experimental data sets. Therefore in here MRA was connected to a maximum-likelihood statistics and a greedy hill-climbing model selection routine. In order to assess the capacities of the so called ML MS MRA, the performance was tested on *in silico* data sets.

First, the network features were systematically varied. It was found that the sensitivity but not precision of the algorithm is affected by noise, therefore experimental noise levels should be controlled for a more concise recapitulation of the network. A reduced noise influence was argued to be achieved either by more precise or repeated measurements. Furthermore, it was found that stronger perturbations result in improved predictions. This is interesting, as MRA relies on linearisation of the underlying dynamical system and is therefore assumed to be only valid for small perturbations. Variation of non-linearity in the system, represented by cooperativity, did not greatly alter performance. Therefore non-linear interaction terms can be well approximated by linear terms, when in steady state. In contrast to this, network size was shown to be negatively correlated with overall performance whereas increasing network density only affected sensitivity. The decrease in performance in cases where a higher number of links is to be expected is possibly caused by the utilised greedy hill-climbing procedure. Therefore ML MS MRA is only suitable to investigate small and sparse networks but can not serve as a genome wide network reconstruction alternative.

Second, it could be shown that ML MS MRA deals well with data, where the global response matrix is not square. Such data may be important in the presence of protein-protein interactions, as protein levels themselves are difficult to perturb without changing the mRNA level. However, it was also shown that this improved performance is only effective if the additional information does not strongly increase model complexity. Not only does the added protein information contribute to improved performance also from the experimental point of view better model descriptions are to be expected. For MRA to function, the perturbation as well as interactions have to be reliably estimated and apparent in the measured data. As demonstrated on posttranslational networks those interactions can sometimes not be asserted on RNA level. This is also true for certain perturbations in genetic networks. Perturbations mediated by siRNA can be accomplished in two ways: accelerated mRNA decay or translational inhibition (Valencia-Sanchez et al., 2006). Often MRA-suitable experiments represent microarray or RNAseq data predicting the activity from the RNA abundance. However if the RNA interference only acts on the translational inhibition (as for example the knockdown of TAZ in Azzolin et al. (2012)), then the mRNA level is unchanged but the corresponding protein is reduced. Thus incorporating additional layers of regulation might be beneficial for both interaction recovery as well as correct perturbation quantification.

Third, the newly proposed methodology was compared to established steady state prediction methods. It could be shown that ML MS MRA can handle different interaction types and additional data better than the established methods MC MRA, plain MRA and NIR. This is particularly interesting as NIR has been considered the best algorithm to deal with perturbation data, or to at least perform similarly well as any other competing algorithm (Camacho et al., 2007; Brazhnik, 2005; Stark et al., 2003). It further was shown that ML MS MRA performs better also on RNA data for sparse, small or nonlinear parameters.

Thus the goal of making MRA a more versatile method seems to be achieved. However, this versatility comes at a cost. For certain beneficial settings in the prediction of transcriptional networks from RNA data, other methods excel. In the test sets as well as on artificial networks from the DREAM Challenge it could be shown that MC MRA might be on average the better choice if a beneficial signal-to-noise ratio is to be expected. MC MRA seems to have inherited this property from the original MRA which under those settings also produced better results when comparing the same number of links that have been recovered by ML MS MRA (not shown). In contrast to plain MRA which can only rank links according to strength of r, MC MRA also provides a confidence level which could be shown to produce similar results than by taking the ML MS MRA cutoff (cf. Suppl. Tab. A.1). Therefore MC MRA is a valid alternative that has to be considered for those experiments that aim to predict small networks from RNA data with strong perturbations and little noise. Depending on the knowledge of environmental parameters and experimental settings the best extension of MRA should be selected. However, it should also be noted that the Levenberg-Marquardt algorithm used in ML MS MRA produces fast and reproducible results whereas the Monte-Carlo approach is not as robust for low sampling rates and is computationally more intense for sufficiently large sampling rates. Further MC MRA seems to be less flexible than ML MS MRA as predictions from additional protein data were mainly worse than ML MS MRA.

Thus, it can be concluded that the described algorithm can be ranked among the best available methods to infer direct regulatory relations from steady states. However, the parameter benchmarking did indicate that in order to successfully apply ML MS MRA to experimental data sets, certain conditions have to be controlled for. Those conditions will be highlighted on a cancer-related reverse engineering application.

2.5.1 Top-down analysis of an oncogenic KRAS-controlled transcriptional network

One of the most crucial tasks in cancer research is to elucidate the changes in wiring that occur upon oncogene emergence. Knowledge about this will help in describing the phenotype that is caused by the oncogene as well as in devising potent therapies. One of the oncogenes that currently receives increasing attention due to the lack of treatment option is the small GTPase RAS. In this effort Stelniec-Klotz et al. (2012) decided to apply MRA to dissolve the regulatory structures that lie underneath the oncogenic transformation of ovarian cancer cells upon KRAS mutation.

Pre-neoplastic rat ovarian surface epithelium (ROSE) cells undergo oncogenic transformation when transfected with mutated KRAS. The transcriptional profile of this transformation has been described previously in Tchernitsa et al. (2004). From the 51 transcription factors that were found to be upregulated, the interaction between seven of those was investigated by knockdown experiments and subsequent RNA and protein measurements. In agreement with the findings of the *in silico* network results the most suitable MRA extension ML MS MRA was applied. From this analysis a hierarchical structure could be discovered with three transcription factors acting upstream of the other four (Fig. 2.10B). Furthermore, a functional assignment could be established by discovering that only the upstream factors are responsible for anchorage-dependent growth (2-D growth).

From the 25 interactions predicted by ML MS MRA at least 15 were confirmed by independent literature findings whereas only one false positive interaction could

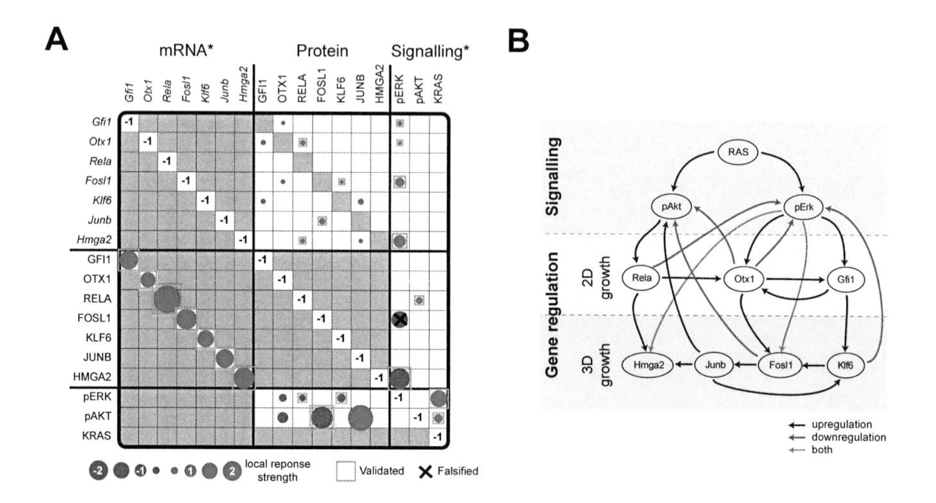

Figure 2.10 MRA model of network downstream of oncogenic KRAS (A) Model fitted local response matrix of the signalling and transcriptional network downstream of oncogenic KRAS in rat ovarian cancer cells. Asterisks denote directly perturbed nodes: mRNAs via siRNA and signalling molecules either by small molecule inhibitors (pERK and pAKT) or ectopical expression of oncogenic KRAS. Measurements were conducted by microarray and qRT-PCR (mRNA) or Western Blot experiments (total- and phospho-protein). Entries surrounded by squares were confirmed and those crossed out falsified by experiments and literature findings. (**B**) Corresponding hierarchical network structure in connection to functional implications. *Data generated by Iwona Stelniec-Klotz and modelling was executed by Nils Blüthgen and Stefan Legewie*

be ascertained (cf. Fig. 2.10A). When reviewing the results garnered from the benchmarking section this exceeds most settings tested in there. The reason for this improved performance seems to reside in the beneficial experimental setup. First, the knockdown efficiency was attempted to be as high as possible ranging between 77-99%, which is beneficial for both sensitivity as well as precision (cf. Fig. 2.4). Second, a fixed error on R (0.4) was assumed which most likely exceeds the real data error for most measurements. This conservative error model might reduce the number of false positives (cf. Fig. 2.5A+B). Third, by letting the algorithm run on replicate measurements and providing protein data but prohibiting protein-protein interactions, the information gain was higher than the increase in possible parameter space (white area in Fig. 2.10A). Thus an overdetermined situation was likely to be created.

The study could also exploit the flexibility that ML MS MRA has conferred to MRA. Next to the knockdown experiments also three signalling readouts and perturbations could be incorporated into the model. This provides one of the first examples that successfully generated a hybrid model connecting gene-regulatory networks to signalling. Furthermore the reported false positive connection - a negative effect of phosphorylated ERK on FOS-like antigen 1 (Fosl1) protein level - proved to be an interesting entry point for further studies. With the help of the model other candidates could be identified by prohibiting this false relationship and allowing the other transcription factors to regulate the protein level of Fosl1. This lead to the prediction that the transcription factor Otx1 might post-transcriptionally regulate Fosl1 which was confirmed in follow-up experiments. Thus, even false model predictions can lead to further insights, if identified and appropriately followed up.

2.5.2 Future directives to advance in genetic network research

Many genetic networks alike signalling networks are structures that evolved to be utilised flexibly in various cell types and under various conditions. Therefore even a model that contains all possible reactions and parameters will be of little use to predict the state that is present in a particular cell when not provided with cell specific information. In cancer this is even aggravated as a general classification is often obstructed by the heterogeneity induced by mutations. Therefore in my opinion network parametrisation from experimental data is an ongoing challenge in genetic network research.

The most unbiased approach has been discussed in this chapter by trying to reverse engineer networks entirely from perturbation data. The here introduced methods that evaluate steady state data have unfortunately shown comparatively poor performances for many parameter settings limiting their general use. Especially the low precision is of concern. As no tested method excelled, the overall information content of the steady state data produced in the *in silico* networks seems to be insufficient to produce better predictions for biologically feasible conditions. From the DREAM challenge contest it was also apparent that other approaches with access to more information can produce more precise and complete ranking predictions. This additional information was probably contained in the more data-extensive time series (550 vs 150 data points for steady state). In order to increase the information content one could either additionally measure specific nodes (i.e. protein) or do repeat measurements to more accurately determine the steady states. Thus, more data points have to be generated which was the original criteria to exclude the time series prediction methods. In the light of the found results a re-evaluation of the currently best available reverse engineering methods from steady state and time series data might be appropriate in order to determine the most information efficient approach. A fair testing environment would be represented by a benchmarking on an equal number of data points derived from identical perturbations of the same network. These tests might help to determine the optimal balance between data requirements and network reliance as well as to single out the most efficient algorithm. From those further tests one might be able to determine whether unbiased reverse engineering methods are experimentally feasible or too cost-intensive to be justified as a standard approach.

An alternative approach to *de novo* network reconstruction consists of a parametrisation of a transcription factor network whose structure has been extracted from available literature data. This approach would require less new experiments but relies on the reliability and accuracy of the provided prior knowledge. When searching for a competent resource for the human genetic interactome it was apparent that available information in existing databases such as TRANSFAC (Matys et al., 2006), ORegAnno (Griffith et al., 2008) or Human Transcriptional Regulation Interaction (HTRI) (Bovolenta et al., 2012) only represent a small fraction of the published literature knowledge. In order to increase this knowledge we manually curated publications that have been pre-filtered by a specifically developed text-mining workflow that weighted abstracts for transcription factor interactions (Thomas et al., 2015). We mainly concentrated on low and medium throughput experiments recording the three evidences physical promoter binding, effect on promoter activity, and co-regulation. Thus the number of false positives is likely to be much smaller than when collecting data from high-throughput experiments. The compiled database called FasTForward DNA[6] has increased the previous knowledge stored in databases by 60% and includes 815 connections between ≈ 370 of the currently known 1391 transcriptional regulators (Vaquerizas et al., 2009). This information is gathered from cells with various origin and differentiation status which can be accounted for by ML MS MRA by pruning unnecessary links. However, due to the research bias only parts of the transcription factor interactome are covered so far. Where sufficient information is present, the data base can be taken as base for the parametrisation of subsets of TF networks to investigate the wiring and strengths in the investigated cellular context. Furthermore, by providing a more comprehensive biological gold standard, the improvement and further development of prediction methods also on larger scale may be facilitated.

Next to resolving the genetic interactome, another critical point in understanding intracellular decision making is the connection of the genetic to other layers of regulations, e.g. the signalling or metabolic network but also other posttran-

[6] `http://fastforward.sys-bio.net/`

scriptional layers. Thereby it is not sufficient to derive them as separate modules that can be plugged into each other. The aspect of dense interlacing has been nicely illustrated for signalling and genetic networks on the relationship between the kinase ERK and the transcription factors c-Fos and FRA1. ERK is reported to activate the transcription factor of c-Fos and is at the same time responsible for the phosphorylation of the newly translated protein to escape rapid degradation (Murphy et al., 2002). Only the stabilised c-FOS can induce transcription. Among the induced targets is the transcription factor FRA1 which is also stabilised by ERK phosphorylation (Vial and Marshall, 2003; Casalino et al., 2003). Thus c-Fos and FRA1 serve as temporal integrators of ERK activity requiring long-term activation of ERK to form a feedforward loop structure (Murphy et al., 2004). Understanding these aspects for more cases would serve as the ultimate link between signalling and genetic networks, strongly strengthening our understanding of the processes happening in the cell. Therefore more modelling studies encompassing crosstalk between other regulatory layers and genetic network members, such as the illustrated oncogenic KRAS network study, are of need.

2.5.3 Improvements for further use of extended MRA

The results from this chapter have demonstrated that ML MS MRA has enabled MRA to be applicable on a more wider range of data sources, interaction types and parameter settings. However the downside of this methodology is that it tends to lose precision with increasing number of recoverable links. In order to increase robustness of model results improvements are to be found.

One aspect was to test out different realisations of the global response coefficient. The theoretical definition ($R = \frac{d\ln(x_{(1)})}{d\ln(x_{(0)})}$ (Kholodenko et al., 1997)) is experimentally approximated by two propositions. The first proposition estimates the global response coefficients from the pre- and post-perturbed steady states by calculating the log fold change and the second by their difference divided by the mean, which was used in this chapter (cf. Eq.1.3.2). The two approximations yield very similar results for $|R_{ij}| \leq 1$ but then deviate as the latter converges to ± 2 whereas the log ratio is not bounded (see Suppl. Fig. A.7). A direct assessment of the two realisations is shown for the benchmarking in Suppl. Fig. A.8. From there it is apparent that while the number of recovered links, i.e. sensitivity, is largely indifferent to the definition of R, the overall performance seems to be better and never worse for the log variant. This seems to predominantly stem from an improved precision. Therefore the log variant appears to produce qualitatively better results than the previously used variant and will be used in the following applications of MRA.

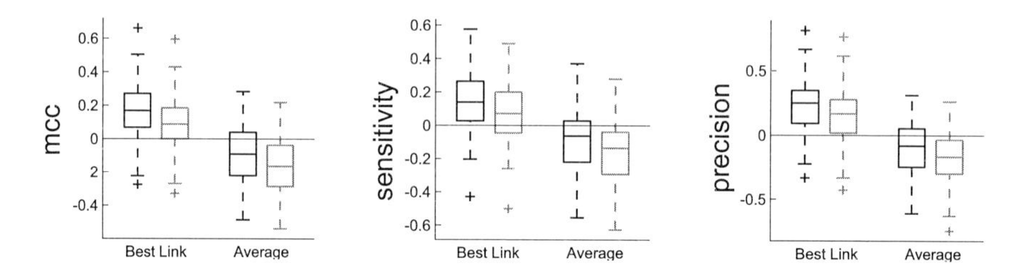

Figure 2.11 Benefit of prior knowledge Boxplots depict performance improvement of ML MS MRA when provided with a network containing the translation and one correct link for the transcriptional network with default parameters. The difference is calculated either by deducting the statistics from scratch (black Boxplot) or from scratch plus an additional true positive (grey Boxplot). Two scenarios are depicted once the link that produced the highest statistics (best link) and once for the average.

This improvement will most likely also apply to the rivalling concepts of ML MS MRA.

Another aspect that could allow MRA to increase performance is to improve the model selection. Two aspects in here hide potential pitfalls. First, the Levenberg-Marquardt routine used for parametrisation in each attempt to alter the network structure represents a local optimiser whose results are dependent on the starting values. Second, the heuristic of greedy hill-climbing itself that is installed to overcome the problem of combinatorial explosion might result in suboptimal wiring realisations as well. For example repeating the model fit by adding simultaneously two or more links might produce a different network. For the starting value problem, sampling of initial values might represent a solution. However, given the numerous fitting steps during network construction this approach might be computationally infeasible for even medium-sized networks. A solution for the suboptimal network structure problem might be in providing prior knowledge[7]. Fig. 2.11 shows the improvement that can be gained when providing ML MS MRA a starting structure with one correct link and the translation (which is always known and is disregarded by the evaluation scheme). It can be seen that the particular link that allowed the best prediction is in more than 75% of cases better than the network calculated from scratch (left black boxplots). This is only slightly lower when correcting the scratch network for the additional correct link in the statistics (left grey boxplots). This demonstrates that even in small networks suboptimal results commonly occur when using this workflow. However, it should be noted that not all link additions lead to better predictions. When comparing the average

[7]An alternative solution for this problem will be discussed in Section 5.2 on page 103.

difference of the 18 possible single link additions to the unbiased approach, the majority of test sets is predicted worse. Therefore a specific order is required in order to arrive at better network solutions which can not be known beforehand.

This dependency on link order is most likely more critical in the extension phase than in the reduction phase. In the reduction phase only changes are accepted that contribute least to the fit. Thus in many cases the error landscape is not expected to change by the removal of superfluous links. In contrast to this, in the extension phase only links that significantly improve the performance are accepted which in the process constantly cause a reshaping of the error landscape.

In conclusion, when the basic wiring is known and the only interest relies in the parametrisation of the network, a compromise solution would be to start with a literature-backed network and to approximate the globally optimal parametrisation by, e.g. re-running the local optimiser with different starting conditions until convergence. After the optimal solution is determined, the network can be pruned off superfluous links in the reduction phase with the starting values being re-cycled from the globally approximated fit. An example of this approach will be presented on a signalling network case in the following chapter.

Modelling EGFR signalling in a colon cancer panel

This work has partly been published in Klinger et al. (2013). Contributions of co-authors are indicated. For details on the experimental methods refer to the materials and methods section of the original article.

Synopsis

The epidermal growth factor receptor (EGFR) signalling network is activated in most solid tumours, and small-molecule drugs targeting this network are increasingly available. However, often only specific combinations of inhibitors are effective. Therefore, the prediction of potent combinatorial treatments is a major challenge in targeted cancer therapy. In this chapter an extended version of modular response analysis is developed and successfully applied to reverse engineer the EGFR network in cancer cells. For this purpose a perturbation data set was devised to monitor the response of RAS/PI3K signalling to stimulations and inhibitions in a panel of colorectal cancer cell lines. By analysing the data with MRA, a negative feedback involving EGFR was detected that mediates strong crosstalk from ERK to AKT. Consequently, when inhibiting ERK, AKT activity is increased in an EGFR-dependent manner. Using the model, it could be predicted that in contrast to single inhibitions, combined inactivation of MEK and EGFR could inactivate both endpoints of RAS, ERK and AKT. The proposed combinatorial inhibition was shown to strongly inhibit growth of BRAF- and KRAS-mutated tumour cells *in vitro* and could be further validated in a KRAS mutant xenograft model.

3.1 Introduction

The signal transduction network downstream of the epidermal growth factor receptor (EGFR) has received much attention, as the majority of human cancers harbours mutations leading to hyper-activation of the associated network (Hanahan and Weinberg, 2011). Based on detailed mechanistic understanding of the network, a large number of targeted therapies has been developed (Herbst et al., 2004; Roberts and Der, 2007; Prenen et al., 2010). However, despite positive treatment responses in some patients, a large fraction of patients do not benefit even if molecular markers such as KRAS or BRAF mutation status are used to stratify patient groups (Karapetis et al., 2008; Walther et al., 2009; Roth et al., 2010). One reason for the somewhat disappointing response rate to these therapies is that they have been developed using the concept of linear signalling pathways downstream of the receptor. However, the EGFR signal is propagated through a complex network (Bublil and Yarden, 2007), involving crosstalks to parallel pathways (Porter and Vaillancourt, 1998) and strong feedback loops on different levels (Blüthgen and Legewie, 2008; Legewie et al., 2008; Cirit et al., 2010; Avraham and Yarden, 2011). Quantitative analysis of these regulatory principles suggested that strong feedbacks can neutralise drug treatment (Friday et al., 2008; Cirit et al., 2010; Sturm et al., 2010; Fritsche-Guenther et al., 2011).

Mathematical modelling of signalling networks can help to understand the behaviour of these complex networks, and can be used to simulate the effect of drugs in such a network. The structure of these mathematical models can be directly deduced from pathway maps (Oda et al., 2005). Detailed mechanistic models based on ordinary differential equations (ODE) have been developed for the EGFR signalling network (Kholodenko et al., 1999; Schoeberl et al., 2002; Nelander et al., 2008). However, for such detailed models the parametrisation remains a major challenge. More coarse-grain modelling approaches, such as logical models or non-mechanistic statistical models require less data for parametrisation (Kreeger et al., 2009; Morris et al., 2011; Saez-Rodriguez et al., 2011, 2009; Tentner et al., 2012). These approaches allow qualitative predictions, but typically fail to deal with feedback loops or do not provide mechanistic insights.

Modular response analysis on the other hand - although requiring similarly small data quantities as coarse grain models - still allows to quantitatively analyse feedback regulation and feedforward loops as well as crosstalks. Being able to monitor these effects is of major interest as these network motifs have major effects on drug sensitivity and network behaviour (Friday et al., 2008; Cirit et al., 2010; Sturm et al., 2010; Fritsche-Guenther et al., 2011). Therefore an experimental setting was devised that best suits the requirements of MRA.

In this chapter, a panel of colon cancer cell lines was exposed to different stimuli and pharmaceutical inhibitors, whereupon key signalling molecules were measured in a medium-throughput approach. The focus of this work was put on

RAS-mediated signal transduction pathways, as they are currently in the strategic focus of targeted therapeutics in solid cancers. To parametrise MRA-based models the existing methodology had to be extended to comply with the signalling data requirements. From the semi-quantitative maps of the wiring between signalling molecules that were calculated by the newly developed model, feedbacks and crosstalks of therapeutic relevance could be identified. The model predicted that EGFR-directed therapeutics might be effective even in tumours carrying a mutation in RAS, if they are provided in combination with RAF or MEK inhibitors. This prediction was confirmed by phenotypic assays and a xenograft model.

3.1.1 Study design

In order to dissect signalling networks in cancer cell lines a combined experimental and theoretical approach was developed to generate predictive mathematical models for their signal transduction pathways (Fig. 3.1A). For a systematic perturbation screen the six colon cancer cell lines LIM1215, HCT116, SW403, SW480, HT29 and RKO were chosen. By targeted sequencing of 46 genes it was consolidated that these cells represent a panel that reflects the genetic diversity of colon cancer harbouring KRAS, BRAF, PI3K or p53 mutations (Table 3.1). This is corroborated by the fact that the cancer incidence of bearing at least one of the

Table 3.1 Cell line panel mutation spectrum List of non-silent single nucleotide polymorphisms (SNPs) detected by sequencing of known mutated regions in 46 cancer-related genes. Asterisks denote mutations reported in the Cancer Genome Project and homozygous mutations are marked by ^{H}. The last column shows the number of patients harbouring the respective mutation in The Cancer Genome Atlas (TCGA). *Raw data generated by Dirk Schumacher and analysed by Pawel Durek*

Gene Symbol	LIM1215	HCT116	SW403	SW480	HT29	RKO	TCGA patients
ABL1	P309A	Y257C					-
APC		K1462R		A1457T/K1462R			-
BRAF					V600E^*	V600E^*	13
CTNNB1	T41A^H						-
FGFR3		S400R		S400R			-
KRAS	A146T	G13D^*	G12V^{H*}	G12V^{H*}			6,8,28
PIK3CA		H1047R^*				H1047R^*	4
SMAD4					Q311X^{H*}		-
SMO		V404M					-
STK11		G58S		G58S			-
TP53			R273H^H	R273H^H	R273H^{H*}		26

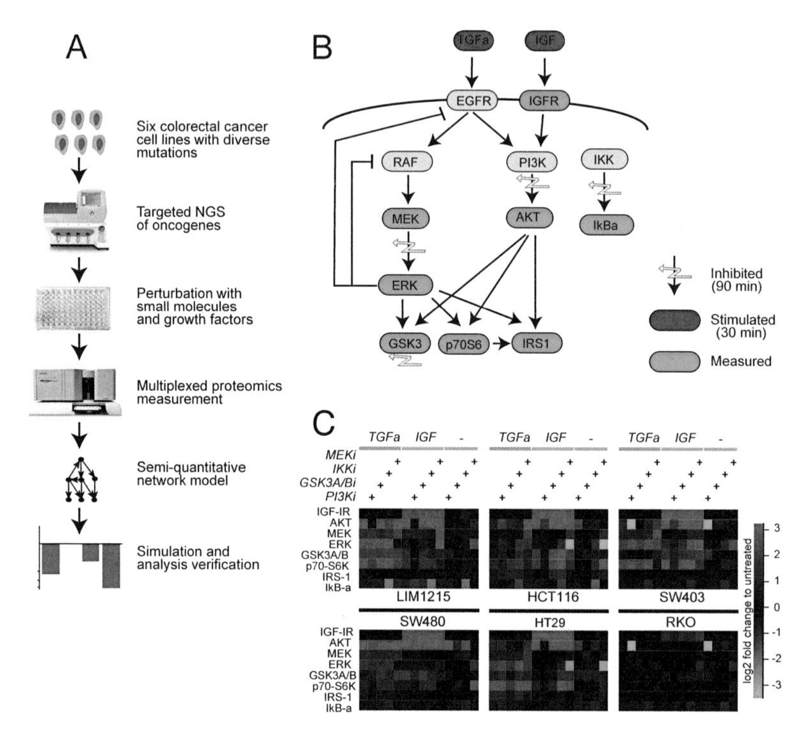

Figure 3.1 Signalling study design (A) General outline of the study. A panel of six colon cancer cell lines was profiled by sequencing on selected cancer-related genes. The cells were then systematically perturbed and phosphorylation of key signalling proteins was measured using the Luminex platform. These data were used for parametrising a mathematical model, which was then used to predict combinatorial treatments. **(B)** Perturbations were realised by two ligands (red nodes) and four pharmacological inhibitors (yellow flashes). These were applied alone and in inhibitor-ligand combinations for the indicated time points. Then eight phosphorylation signals were measured (blue nodes). **(C)** Heatmap containing the corresponding log2 fold changes to solvent control. Displayed response range is limited to ± 3.5 ($\approx$10-fold). *Raw data for* ***C*** *generated by Anja Sieber and Cornelia Gieseler.*

mutations in Table 3.1 was $\geq 94\%$ in all four different colon cancer cohort studies included in cBioPortal (Gao et al., 2013) as of September 2015.

To generate a semi-quantitative database for further mathematical modelling, phosphorylation of selected signalling molecules was measured for network perturbations. In line with previous approaches (Nelander et al., 2008; Saez-Rodriguez et al., 2009, 2011; Morris et al., 2011), a pair-wise design of perturbations was chosen, where each inhibition was combined with each stimulus. Specifically, the cells were stimulated with two growth factors, TGFα and IGF, activating the EGF receptor and the IGF receptor, respectively (shown in red in Fig. 3.1B). The cells have been pre-incubated for 1 h with pharmacological inhibitors against the kinases MEK, PI3K, IKK or GSK-3α/β (flashes in Fig. 3.1B). As ligand stimulations typically display a strong transient response followed by a long-term plateau, time-series experiments were performed to determine optimal time points for the experiments (see Suppl. Fig. B.1). It was found that responses to TGFα as well as IGF peaked within the first 10 minutes, and reached a plateau after 30 minutes. Thus, it was decided to take the 30-minutes time point for stimulations, to best approximate the steady state requirements of MRA. Then the phosphorylation of eight key signalling proteins (AKT^{S473}, ERK2$^{T185/Y187}$, MEK1$^{S217/S221}$, p70S6K$^{T421/S424}$, IGF-IR^{Y1131}, GSK-3$\alpha/\beta^{S21/S9}$, IkB-$\alpha^{S32/S36}$, and IRS-1$^{S636/S639}$) that are within or in close proximity to the stimulated receptors or inhibited kinases (shown in blue in Fig. 3.1B) were measured simultaneously using antibody-coated bead technology. Due to the design of this screen replicates would have been too costly. Therefore to be able to generate a representative error model, a subset of experiments was measured in replicates, namely the unperturbed state and all single inhibition measurements.

The resulting data sets are shown in Fig. 3.1C as log2 fold changes compared to the average unperturbed controls. When comparing the six different cell lines, strong response patterns to the perturbation can be perceived in all cells with the exception of RKO cells. This cell line only showed a clear response for few signals and was excluded from further analysis since the few reactive signals only reflected the known facts that AKT is downstream of PI3K and IGFR is activated by IGF. The remaining five cell lines still represent colon cancer well enough as the two hot-spot mutations found in RKO were also present in HCT116 and HT29 cells (see Table 3.1). All other cell lines, while clearly responding to the inhibitors and stimuli, showed subtle differences in their response patterns. For example, IRS-1 phosphorylation changed strongly in HCT116 cells for specific treatments, but was unaltered in SW403 cells. To further infer qualitative and quantitative differences in the interactions between the signalling nodes and to predict effects of combinatorial treatments, the data was now to be analysed by an appropriate modelling approach of MRA.

3.2 Adjusting MRA to model signalling data

Previously, data generated for MRA consisted of systematic application of single perturbations. The aim of this chapter, however, was to investigate the network behaviour in response to more than one perturbation and to generate predictions for suitable treatment combinations. Therefore MRA had to be adjusted in order to cope with multiple perturbations, including opposing perturbations such as stimuli and inhibitions. The modular response analysis as introduced by Kholodenko and colleagues was based on the assumption that all signals can be perturbed and measured. If this is not the case the network has to be condensed to modules until the requirement is met since a full matrix of **R** is required for inversion. However, in this data set some perturbed nodes could not be measured and some measured nodes could not be perturbed (Fig. 3.1B). In addition, modularisation can not be applied in structures like the branching from RAS to RAF and PI3K where all three molecules were not measured and only PI3K was perturbed. This will introduce parameters that can not be estimated alone but only in combination. These so-called structural non-identifiabilities can cause the log likelihood ratio test to be not reliably approximated by a χ^2 distribution (Timmer et al., 2004). In addition structural non-identifiabilities can also lead to meaningless estimations for single parameters. Therefore solutions for the following four points had to be developed:

1. Develop an approach to handle incomplete information on perturbation and measurement,
2. Integrate ligand stimulation and small molecule inhibition as perturbations - also in combination,
3. Detect non-identifiable parameters and replace them with identifiable combinations,
4. Adapt the model selection strategy introduced in the previous chapter to the new circumstances created by the above points.

The found solutions to these four problems will be explained and illustrated in the following sections.

3.2.1 Modelling incomplete data

The original equation of MRA states that the global response matrix **R** equals the negative inverted local response matrix **r** times the initial perturbations **p** (cf. Eq. 1.4 on page 10). When demanding the rows of $\mathbf{r}^{-1}$ to be normalised by the

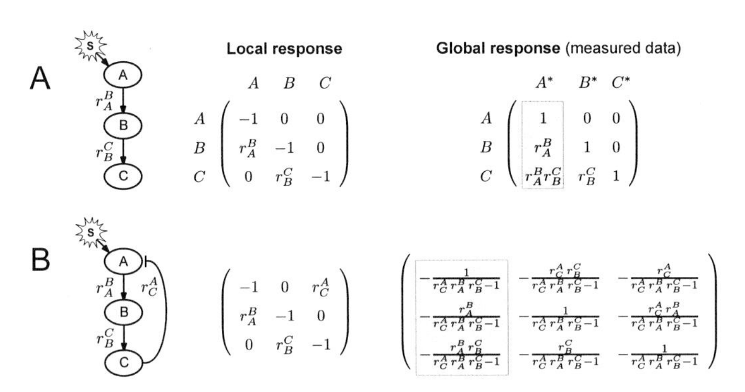

Figure 3.2 Symbolic entries in response matrices of MRA (A) The left graph shows a linear network consisting of three nodes. In the middle the weights of the network are contained in the local response matrix **r** with ingoing nodes indicated in the column and outgoing nodes indexed in the row. On the right the observable response of the network, i.e. global response matrix, is shown. In this simple network only the effect of perturbation A^* on C is not executed directly allowing an intuitive derivation of the underlying local response matrix, i.e. multiply along the path. **(B)** Depicts the changes in local and global response matrix when a feedback from C to A is added to the network from (A). Note when only perturbing at node A (indicated by stimulus S) only the first column of **R** can be measured (highlighted). Asterisks denote directly perturbed nodes with perturbation strengths set to 1.

corresponding perturbation entries of **p**, the right hand side of the equation can be reduced such that it only depends on the local response coefficients

$$\mathbf{R} = -\mathbf{r}^{-1}\mathbf{p} \quad \overset{p\in\{0,1\}}{\Longleftrightarrow} \quad \mathbf{R} = -\mathbf{r}^{-1} \quad . \tag{3.1}$$

In Fig. 3.2 the corresponding symbolic relationship of local and global response matrix is exemplified on a three-tiered network for a linear cascade with and without feedback. This feedback now causes the originally sparse global response matrix to be fully connected although only half of the entries in **r** are non-zero. As a consequence, perturbations targeting any node will propagate through the entire network. As in this work only a subset of the global response matrix can be measured (exemplified as highlighted area in Fig. 3.2) inversion of **R** to arrive at **r** is not feasible. Instead, models for the single perturbations and measurements, i.e. single entries of **R**), have to be devised and fitted to the measured data.

Thus the global response of node k to the perturbation in node j can be formulated as:

$$R_j^k = (-\mathbf{r}^{-1})^k \, \mathbf{p}_j^* = \sum_{i=1}^{N} (-\mathbf{r}^{-1})_i^k p_j^{i\,*} \quad , \tag{3.2}$$

where $\mathbf{p}_j^*$ is a binary vector that only contains a non-zero entry at the position of the directly perturbed node j. The resulting formulas for each measurable global response coefficient are then taken as the objective function that will be fitted to the experimental measurements. It is of note that via this entry-wise evaluation scheme it has now become possible to model simultaneous perturbations, i.e. a binary perturbation vector with more than one non-zero entry. This new property is essential to model and predict combinatorial treatments.

3.2.2 Integration of different perturbation types

Starting from the entry-wise fitting framework, the effect of the two principal perturbation types applied on the signalling data set can now be defined: (i) ligand-driven stimulation and (ii) small-molecule-mediated inhibition (cf. Fig. 3.3).

Stimulation of node j by a ligand S could in principle be simulated by encoding the strength in vector $\mathbf{p}_j$ as in Eq. 3.2 with the jth entry representing the receptor node. However, for the partial MRA approach it proved advisable to keep perturbation vectors binary and to transfer any non-binary information to the local response matrix. Consequently, ligands were classified as additional nodes in the response coefficient matrices and the corresponding entry in the perturbation vector p_j^i was set to one. The actual strength of the stimulation was then placed into $\mathbf{r}$ to yield an estimate of the induced fold change at the receptor level. This is illustrated in the example in Fig. 3.3, where the perturbation strength of stimulus S is encoded in the response coefficient r_S^A.

In contrast to ligands, kinase inhibitors do not create new signals but negatively affect the innate signal sent out by the targeted molecule. This effect can be modelled by a negative perturbation of the targeted node m by an inhibitor $i_m \in (-\infty, 0]$. In Fig. 3.3 it is shown that the inhibition on node B is not apparent at the global response of B, but at the downstream node C whose response is altered by the inhibitory term. Inhibitors, unlike ligands, act within the signalling cascade and do not necessarily impose the only rate changing perturbation on the inhibited node as perturbations placed upstream can act on that node as well. One approach to include the inhibitory strength into $\mathbf{r}$ would be to define a new node that is placed between the inhibited node and all downstream targets. However, the value of that coefficient has to switch between the inhibitory strength in the presence of the inhibitor and one as neutral relay in the absence of the inhibitor. Therefore incorporation of the inhibition strength in the local response matrix can

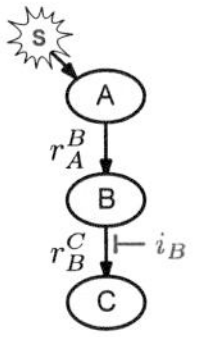

S A r_A^B B r_B^C i_B C

$$\mathbf{r}(\tilde{\mathbf{r}}) = \begin{array}{c} \\ S \\ A \\ B \\ C \end{array} \begin{array}{c} \begin{array}{cccc} S & A & B & C \end{array} \\ \begin{pmatrix} -1 & 0 & 0 & 0 \\ r_S^A & -1 & 0 & 0 \\ 0 & r_A^B & -1 & 0 \\ 0 & 0 & (e^{i_B})r_B^C & -1 \end{pmatrix} \end{array}$$

$$\mathbf{R} = \begin{array}{c} \\ S \\ A \\ B \\ C \end{array} \begin{array}{c} \begin{array}{ccc} S^* & B^* & S^*B^* \end{array} \\ \begin{pmatrix} 1 & 0 & 1 \\ r_S^A & 0 & r_S^A \\ r_S^A r_A^B & 1 & r_S^A r_A^B \\ r_S^A r_A^B r_B^C & i_B r_B^C & r_S^A r_A^B e^{i_B} r_B^C + i_B r_B^C \end{pmatrix} \end{array}$$

Figure 3.3 Integration of (simultaneous) stimulation and inhibition in MRA (**left**) The same network is chosen as in Fig. 3.2A with the only difference that next to a stimulation by ligand S, node B was targeted by an inhibitor i_B (marked in red). (**middle**) Local response matrix which contains an additional node S whose only non-zero coefficient r_S^A encodes the perturbation strength of stimulus S. The bracketed red entry denotes the inhibitor term in the modified $\tilde{\mathbf{r}}$ that is multiplied to all outgoing coefficients of the inhibited node to dampen the upstream stimulus. The actual inhibitory parameter $i_B \in (-\infty, 0]$ is for now encoded in a separate perturbation matrix $\mathbf{i}$. (**right**) Global responses for single stimulation, inhibition and a combination thereof. The responses to single perturbations comply with MRA rule. However, when combined the global response of the doubly affected node C contains an extra term e^{i_B} that dampens the stimulus and is exponentially coupled to the inhibitory strength.

not be done in a straight forward manner and remains in a separate perturbation matrix $\mathbf{i}$.

Since the experimental design allows the simultaneous action of ligands and inhibitors, a second effect of the inhibition has to be taken into account as well. Next to lowering the basal activity of the node, i.e. exert a negative fold change on downstream nodes, inhibitors also reduce the ability of the inhibited node to relay upstream signals generated by, e.g. an additional stimulus. This dampening of the incoming signal is modelled by multiplying an inhibitory term $\tilde{i}_m \in [0..1]$ to all outgoing local response coefficients starting at node m. This effect is expressed in a modified matrix $\tilde{\mathbf{r}}$ where each response coefficient placed in the column of the inhibited node gets multiplied by the inhibitory term (see Fig. 3.3). By recognising that the two effects of the inhibitor are coupled in strength one can express both effects with one parameter by modelling an exponential relationship:

$$\tilde{i}_m = \mathrm{e}^{i_m} \quad . \tag{3.3}$$

Thus the response in node k to a combined perturbation consisting of a stimulus at node j and an inhibition at node m can be expressed by

$$R_{j,m}^k = (-\tilde{\mathrm{r}}^{-1})^k \, \mathbf{p}_j^* + (-\mathrm{r}^{-1})^k \, \mathbf{i}_m \quad . \tag{3.4}$$

Thereby the coefficient $\tilde{i}_m$ included in $\tilde{\mathbf{r}}$ dampens the effect of the stimulation on downstream nodes and i_m reduces the basal activity of the affected node (cf. Fig. 3.3 third column in $\mathbf{R}$).

It should be noted that not all repressed nodes are expected to exhibit basal activity, e.g. receptors are usually inactive when not externally stimulated (unless the cell contains mutations that lead to constant production of their ligands). For those inactive receptors the second term in Eq. 3.4 is removed and the inhibitory strength estimation stems solely from modelling the dampening effect.

3.2.3 Removing non-identifiable parameters

Having defined all parameters in the objective functions it is now necessary to determine the identifiable parameter combinations. The first part of this section will explain the chosen approach on a simple subsystem containing only stimulations and afterwards the procedure to incorporate inhibitors is discussed. In the simple case for each perturbation and for each measured node, the model prediction (in terms of log-fold change of the unperturbed steady state) can be written as a function of the negative inverse of matrix $\mathbf{r}$. As already apparent from the examples in Figures 3.2 and 3.3, the equations for the response of the network typically contain products of entries of $\mathbf{r}$ along paths in the network (cf. Kholodenko et al. (1997)). Depending on the experimental design, i.e. which nodes are perturbed or measured, it is often the case that some r's only occur in combination in these paths, never alone.

For example in the network in Fig. 3.4 the internal node B is not measured. Therefore r_A^B occurs only in combination as can be seen on the measurable entries

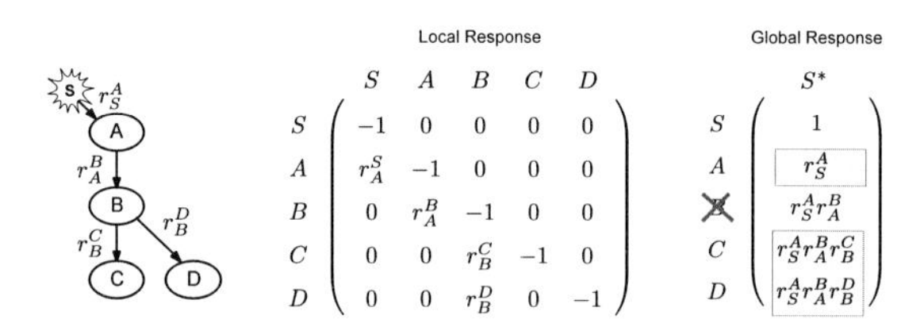

Figure 3.4 Structural non-identifiability (left) Four-node network stimulated at node A. Nodes with highlighted background indicate measured nodes. (**middle**) Local response matrix with agent in the column and recipient in the row. (**right**) Global responses to the stimulus S^* with the stimulus strength included as response coefficient. Due to the global response of node B not being measured the four parameters have to be fitted from three entries, introducing non-identifiabilities.

of the global response matrix (highlighted). Since more parameters than entries in **R** have to be fitted, an infinite number of parameter realisations can be found that give identical solutions. Thus non-identfiability might prolong fitting procedures and can even lead to suboptimal parameter solutions. In addition the fitted values of those non-identifiable parameters are themselves meaningless as they can be only interpreted in combination. One intuitive solution for the illustrated example, would be to re-define a new module AB with one ingoing and outgoing node thereby effectively fitting the identifiable parameter combinations $r_A^B r_B^C$ and $r_A^B r_B^D$.

The MRA approach used in the previous chapter utilised a model selection procedure, where nodes of the network were systematically removed or added. Applying this strategy in here will cause structural non-identifiabilities to occur and disappear. Therefore when a model selection procedure is necessary an automated strategy to detect those structures and to fit only identifiable parameter combinations is needed. In order to derive those combinations, the equation for the global response entries was first re-parameterised in terms of the path strengths P:

$$R_j^k = f(P)_j^k \,. \tag{3.5}$$

$f(P)_j^k$ is the function of paths P_l that lead from node j to k. The functions of paths encode the non multiplicative parts of the response coefficient which occur, e.g. in feedbacks (cf. Fig 3.2). The paths themselves are defined as pure products of r's:

$$P_l = \prod_{(n,m)\,\text{edges in path}\,l} r_n^m \quad . \tag{3.6}$$

By log-transforming Eq. 3.6 the path strength of each P_l can then be written as sums of r's[1] :

$$\log P_l = \sum_{n,m} a_n^m \log r_n^m \,, \tag{3.7}$$

with a_n^m standing for the exponent of r_n^m in path P_l (usually 0 if the path does not go from node n to m, and 1 otherwise). The set of linear equations in Eq. 3.7 can be reformulated in matrix form:

$$0 = [\mathbf{A}, \text{-}\mathbf{I}] \times (\log r_1, \ldots \log r_M, \log P_1, \ldots \log P_N)^T \,, \tag{3.8}$$

where **A** contains the exponents a_n^m and **I** is the identity matrix with size equal to the number of paths. By applying Gaussian elimination to [**A**, -**I**], one arrives at

[1]For negative paths $|P_l|$ is transformed and the sign information is stored in the constructor function $f(P_l)$.

the reduced row echelon matrix $\mathbf{G}$, which can be divided into 4 parts:

$$\mathbf{G} = \left(\begin{array}{c|c} A' & G_1 \\ \hline 0 & G_2 \end{array} \right) . \tag{3.9}$$

The number of rows in A' equals the number of independent parameters. Together with G_1 it provides rules how to construct them. G_2 identifies shared or equal paths.

These entries are now exemplarily calculated for the simple example given in Fig. 3.4. In there matrix $\mathbf{A}$ can be determined from the global responses of stimulus S as

$$\mathbf{A} = \left(\begin{array}{c|cccc} & r_S^A & r_A^B & r_B^C & r_B^D \\ \hline P_1 & 1 & 0 & 0 & 0 \\ P_2 & 1 & 1 & 1 & 0 \\ P_3 & 1 & 1 & 0 & 1 \end{array} \right) . \tag{3.10}$$

After applying the Gaussian elimination algorithm one arrives at:

$$\mathbf{G} = \left(\begin{array}{cccc|ccc} r_S^A & r_A^B & r_B^C & r_B^D & P_1 & P_2 & P_3 \\ \hline 1 & 0 & 0 & 0 & -1 & 0 & 0 \\ 0 & 1 & 0 & 1 & 1 & 0 & -1 \\ 0 & 0 & 1 & -1 & 0 & -1 & 1 \\ \hline & & & & & & \end{array} \right) . \tag{3.11}$$

Since in this example the rank of $\mathbf{A}$ is equal to the number of rows, the resulting $\mathbf{G}$ contains only the upper two parts A' and G_1. A' suggests to re-parameterise the response coefficients into three combinations c:

$$\begin{aligned} P_1 &= c_1 = r_S^A \\ \frac{P_3}{P_1} &= c_2 = r_A^B r_B^D \\ \frac{P_2}{P_3} &= c_3 = \frac{r_B^C}{r_B^D} \quad . \end{aligned}$$

The thus extracted identifiable combinations c can now be taken as parameters to fit the global response matrix to the experimental data where the P_l's in $f(P)$ can be derived from the mapping of c's from G_1. After finishing the fitting procedure the best fitting parametrisation can then be used to reconstruct the identifiable combination of local response coefficients from the c's.

In the present model case stimulatory strengths are already incorporated in the local response matrix. Inhibitory perturbations i_m are treated analogously to the stimuli as response coefficients and are included in matrix $\mathbf{A}$ as additional columns:

$$0 = [\mathbf{A}, -\mathbf{I}] \times (\log r_1, \ldots \log r_M, \log i_1 \ldots \log i_n, \log P_1, \ldots \log P_N)^T \,, \qquad (3.12)$$

For the Gaussian elimination it is neglected which of the two actions (basal inhibition or stimulus attenuation) the inhibitory strength is assigned to in this particular path as it only considers the stoichiometry. However, it should be differentiated in the subsequent fitting process. When including inhibitors, Eq. 3.4 should be therefore re-defined as:

$$\mathbf{R_{model}}_j^k = f(g(c))_j^k \quad , \qquad (3.13)$$

where $g(c)$ represents the constructor function to regenerate the paths from the identifiable combinations of r's and i's.

To summarise, for each measurement the constellation of local response coefficients and acting inhibitors has to be determined. Thereby each product represents one path and the global responses are represented by a function of those paths. The paths themselves are made up of products of local response coefficients, stimuli, and inhibitory perturbations. By using Gaussian elimination on the dependency matrix of the path, identifiable combinations of parameters can be extracted. An example of identifiable parameter combinations described for the HT29 cell model is shown in Suppl. Fig. B.2.

As long as the entries in the global response matrix are composed of pure products, exactly one R_j^k is represented by one P_l as shown in the example in Fig. 3.4. However, through feedbacks (cf. Fig. 3.2B), simultaneous perturbations acting inside one path (cf. Fig. 3.3) or branching paths that are converging again, sums are introduced which will result in one R_j^k being represented by more than one P_l. The experiments should therefore be designed such, that all paths are identifiable. This means, next to combinations of perturbations always the effect of single perturbations should be measured and feedback structures and branch-merge patterns should be broken by perturbations within. Following these rules, non-identifiable paths should not occur during the fitting procedure.

As long as links of the network are only removed, this transformation is valid given the rules were applicable for the starting network. For link additions, occurrence of non-identifiable paths is likely to happen. This problem can be solved by applying an additional non-identifiability analysis on the path level with respect to the measured global responses.

3.2.4 Parameter estimation strategy

With these strategies devised to handle incompletely perturbed and measured signalling data, the methodology for the model selection had to be adjusted. In principal the same likelihood function was employed as for the prediction of genetic networks (cf. Eq. 2.2 on page 19) with two differences. First, not the local response coefficients were fitted but the identifiable combinations c for which construction rules to either r/i or to P can be derived from G. Second, the level of data fitting was shifted to the data side such that instead of the difference in R the perturbed steady states $x_{(1)}$ were used as reference:

$$L(\mathbf{c}|\mathbf{x}_{(1)}) \approx -\chi^2 = -\sum_{k,j=1}^{N} \left(\frac{x_{(1)}{}_j^k - x_{(0)}{}_k \, e^{\mathbf{R}_{\mathbf{model}}{}_j^k}}{\epsilon_j^k} \right)^2 . \tag{3.14}$$

This function was now maximised by fitting the combinations c to the corresponding entries in the measured global response matrix $\mathbf{R}_{\mathbf{model}}$. Additionally, the function served as the cost function to decide which links were kept in the network and which links were superfluous.

The error terms ϵ_j^k used for model fitting were taken from an error model that was derived from replicate measurements and blanks (see Suppl. Sec. B 2).

For three reasons it was decided to start with a prior knowledge-based network which was then subsequently reduced. First, from the previous chapter it became apparent that network reconstruction from scratch produced a comparatively low precision for ML MS MRA. Second, in signalling data sets, due to the limited ability to perturb or measure nodes the information contained in the data is even more scarce than for genetic data sets. And third, addition of links might produce non-identifiable paths which have to be accounted for in the algorithm. For this particular signalling pathway the prior knowledge approach was feasible as it has been extensively studied and the network structure has been largely disclosed.

Alterations of the network structure were considered based on whether links contribute to the fit or not. More specifically, when removing links, it was assumed that the fit in terms of the likelihood worsens due to the reduced model complexity. Since structural non-identifiabilities were removed, a similar procedure could be applied as for the prediction of the genetic networks (cf. Fig. 3.5). In order to avoid local minima for the initial network model, it was sought to obtain an parameter estimate near the global optimum by 10^4 Monte Carlo simulations of the Levenberg-Marquardt fitting procedure. Subsequently, local error minimisation was applied to

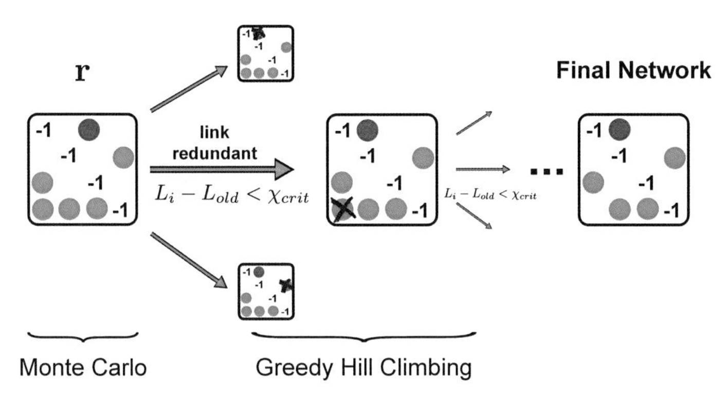

Figure 3.5 Model selection procedure After determining a suitable starting network structure from the literature the non-identifiable parameters are estimated by Levenberg-Marquardt algorithm for 10^4 Monte Carlo simulations to ensure optimal parametrisation. The likelihood of the best fit L_{old} is then compared to the best fit of all network realisations reduced by one link. Thereby the starting parameter values for the new fits are taken from the best fit of the initial network. If the difference is below a threshold ($\alpha = 0.05$ for χ^2 distribution under consideration of the degrees of freedom) the reduced network is taken as the new starting network that is compared to models with yet another link removed. This parameter reduction by the so-called greedy hill-climbing procedure is continued until all $L_i - L_{old} > \chi_{crit}$.

all networks reduced by one link, starting with the parameter set derived from the initial Monte Carlo simulation. The likelihood ratio test was calculated to decide whether the less complex model (i) was not able to fit the data (i.e. $p > 0.05$), or (ii) could be stated as minimal model that is sufficient to explain the data. Thus the links were reduced in a greedy hill-climbing procedure until the network could not be further reduced (see Fig. 3.5). After the final network structure has been determined, identifiable response coefficients and identifiable combinations thereof were calculated from the fitted parameters using the relationship derived from the Gaussian elimination matrix G.

3.3 Modelling results

In Fig. 3.6 all pieces derived for the novel approach to reverse engineer a signalling network are summarised. The model requires two inputs (i) an appropriate systematic perturbation data set with enough replicate measurements to calculate an antibody-wise error and (ii) enough prior knowledge of the underlying network in the literature to derive a starting network that is expected to be in the vicinity of the real connectivity. This proved to be the best strategy to improve the qual-

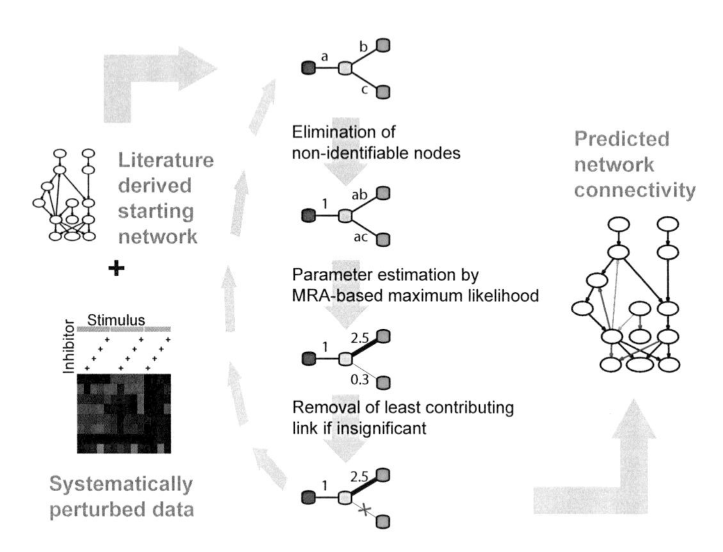

Figure 3.6 MRA Modelling approach of signalling data Systematic perturbation data and a starting network serve as input. The core-fitting routine consists of three steps, and is illustrated by a four-node example network. First, non-identifiable parameters are detected and re-parametrised. Second, the identifiable parameter combinations are fitted to the experimental data. Third, connections not significantly contributing to the fit are removed and subsequently the remaining parameters are refitted. These three steps are iteratively repeated until the minimal model is reached. The resulting network contains information about the strength, direction and sign of its connections.

ity of fits achieved with the local optimisation algorithm (see also Sec. 2.5.3 on page 43). The model as such reduced the network by applying the three modules non-identifiability analysis, parameter fitting and link contribution analysis. The resulting network parameters and connectivities as well as the goodness of fit can now be assessed to study model characteristics.

3.3.1 Model fit suggests extension of the original network model

When applying the modelling procedure to the perturbation data sets of the five selected cell lines, fits were obtained with χ^2 scores between 116 and 470 (Fig. 3.7A, dark bars). The algorithm then pruned the network and could remove several links in each cell line without significant increase in the χ^2 scores (Fig. 3.7A, blue bars), indicating that these links were not important in mediating the perturbed signals (see reduction example in Suppl. Fig. B.2).

When comparing the model predictions with the original data, it was observed that the models based on the original network in Fig. 3.1A failed to explain an increase of MEK and ERK phosphorylation after IGF treatment that was visible in multiple cells (see correlation in Fig. 3.7B). When checking back with literature, it was found that IGF-1 can indeed lead to activated RAS by triggering the activation chain via IGF-IR, IRS-1 and Grb-2 (Florini et al., 1996). Hence, a link from IGF to RAF was included, which improved model performance for all five cell lines (red bars in Figure 3.7A).

Additionally, the model could not account for an increase of the phospho-ERK level upon treatment with IKK inhibitor BMS345541, which was particularly pronounced in HCT116 and SW403 cells. Interestingly, this increase in ERK phosphorylation did not coincide with an increase in MEK levels but was evident in ERK-downstream effector p70S6K (Figure 3.7C). Thus the effect of IKK inhibition on ERK could not be explained via upstream components in the pathway. It was

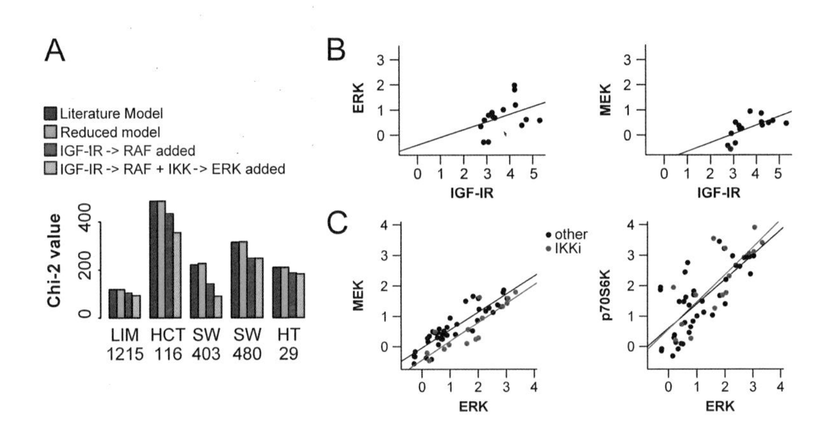

Figure 3.7 Cell-wise scores for different EGFR models Observations in fit and log2 fold change to untreated data for all five cell lines plead for alterations of the original network described in Figure 3.1. **(A)** χ^2 scores of the models trained on data of the five cell lines. Dark bars indicate scores for the initial model. Blue, red and yellow bars show scores after model reduction of literature network, with a link from IGFR to RAF, or when additionally including a link from IKK to ERK, respectively. **(B)** Scatterplot of IGF-IR on ERK and MEK indicates a causal relationship underscored by a Pearson correlation of 0.39 and 0.63, respectively. Subset of data is plotted including all combinations involving IGF-1 treatment excluding interfering inhibitors of MEK and IKK. **(C)** Scatterplot of phosphorylation of ERK-MEK or ERK-P70S6K for treatments involving IKK inhibitor BMS345541 (red) and all other treatments (black). Samples with MEK inhibitor were left out due to the de-correlating effect of the ERK-RAF feedback. Upon IKK inhibitor treatment a shift towards ERK is notable on the MEK-ERK plot whereas no shift occurs for the ERK downstream target p70S6K implicating a direct effect from IKK on ERK.

decided to include this interaction in the model and to repeat the model selection procedure. The inclusion of a link from IKK to ERK improved the fit of the model considerably for HCT116 and SW403 cells (see decrease of χ^2 score in Fig. 3.7A orange bars) and showed no improvement for the other cell lines. However, as this potential link has not been described in the available literature, a more thorough characterisation of this relationship was conducted.

IKK inhibitor effect on ERK is artifactual

The observed increase in ERK phosphorylation after treatment with IKK inhibitors may be either due to unspecific effects of the chosen inhibitor or due to direct or indirect regulation of ERK by IKK or its downstream kinases. To solve this question it was therefore decided to characterise this interaction further. It was found that treatment with the IKK inhibitor BMS345541 resulted in an increase of phosphorylation of ERK within 1 h (Fig. 3.8A, left panel), but treatment with two other IKK inhibitors had no effect on ERK phosphorylation, although they blocked phosphorylation of IkB-α to a similar degree at this time point (Fig. 3.8A, middle panel). This finding suggests that one can dismiss the downstream target IkB-α as the regulator of ERK. Since the two IKK inhibitors PS1145 and PHA408 - that had no effect on ERK - are reported to target the IKK-β isoform (Karin et al., 2004; Mbalaviele et al., 2009) this kinase can be ruled out as well. Whereas both single isoform inhibitors only induced a temporal downregulation of IkB-α that subsided after two to three hours, BMS34551 produced a sustained inhibition even reducing IkB-α phosphorylation further at later time points. One possible explanation for the difference might be that BMS345541 can additionally inhibit IKK-α, albeit with a ten-fold less affinity than to the β-isoform (Burke et al., 2003). If IKK-α is responsible for the observed discrepancy, a mechanism of isoform compensation can be imagined that acts in a similar fashion as the previously described mechanism found for the two ERK isoforms (Fritsche-Guenther et al., 2011).

However, even though the IKK-α isoform may be more activated at later time points, it seems to play no role for early time points as the common target phosphorylation of IkB-α is reduced to the same degree by all three inhibitors one hour post inhibition. Together with the fact that the observed increase of ERK phosphorylation takes place within the first hour of inhibition it is unlikely that ERK could be under the control of IKKα. Thus, it can be concluded that instead of a novel interaction between IKK and ERK the modelling procedure presumably helped to identify an unknown side effect of BMS345541 on ERK.

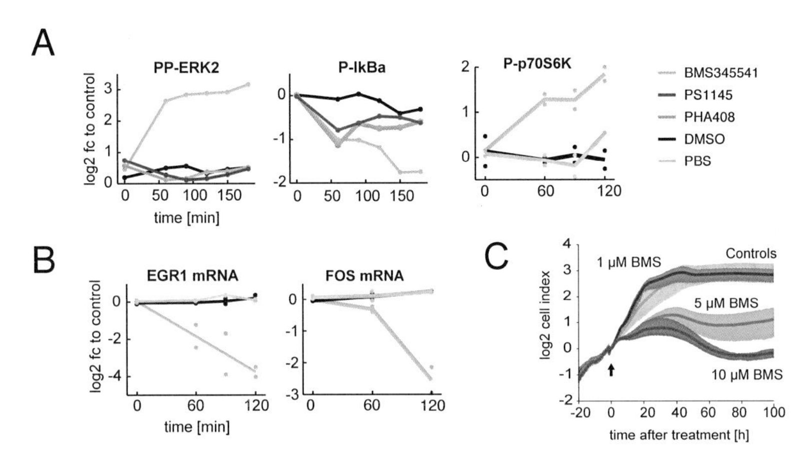

Figure 3.8 Studying the effect of BMS345541 on ERK (A, B) Log2 fold change in time-series experiments after treatment with IKK inhibitors or solvent controls (see color legend) to investigate the IKK-ERK relationship in HCT116 cells. **(A)** IKK inhibitor BMS345541 treatment results in an increased phospho-ERK level, whereas treatment with IKK inhibitors PHA408, PS1145 and solvent control (DMSO) results in no increase. Inhibition of IkB-α phosphorylation is comparable for all three inhibitors after 1 h (Luminex, n=1). Phosphorylation of p70S6K, a cytoplasmic target of ERK, increases after 60 min of BMS345541 treatment (western blot, n=2) when compared with DMSO and PBS control. **(B)** Expression of ERK target genes EGR1 and FOS decreases after BMS345541 treatment (qRT-PCR, n=2). **(C)** xCELLigence measurements of HCT116 cell growth in response to three different concentrations of BMS345541 or its solvent control DMSO (grey) at time 0 (arrow) grown in full serum. Shown is the maximal range of $n \geq 3$ independent wells along with the mean (thick line). Note that for concentration of 10μM cells even reach a lower cell index than at time 0 indicating cell death. *Measurements by Raphaela Fritsche and Figure C generated by Franziska Witzel.*

Interestingly, while phosphorylation of ERK and its cytoplasmic target p70S6K (Lee-Fruman et al., 1999; Lehman and Gomez-Cambronero, 2002) are increased in response to BMS345541 (Fig. 3.8A, right), the nuclear activity of ERK seems to be decreased, as expression of typical immediate-early target genes of ERK such as EGR1 and FOS is strongly reduced (Fig. 3.8B). This suggests that while ERK phosphorylation is increased ERK activity seems to be redirected toward cytoplasmic targets and therefore treatment with BMS345541 may result in the repression of typical genetic programs stimulated by ERK. In line with this hypothesis, it was observed that treatment with moderate concentrations of BMS345541 resulted in strongly impaired proliferation of HCT116 cells (Fig. 3.8C).

The hypothesis that BMS345541 treatment prevents ERK from entering the nucleus could also provide an explanation of the observed increase in phosphorylation.

ERK is described to enter the nucleus by diffusion and is thus controlled by location sequences of binding partners (Burack and Shaw, 2005). MEK on the contrary contains a nuclear export sequence which keeps it out of the core and therefore nuclear ERK is less exposed to MEK phosphorylation (Burack and Shaw, 2005). Supposing that ERK is retained in the cytoplasm, it is constantly exposed to MEK and can therefore reach much higher phosphorylation levels. This putative mechanism could serve as an explanation how the phospho-ERK level can be increased without increasing the active MEK fraction. In spite of this exciting observation the artifactual IKK inhibitor could not be part of the further modelling procedure. Therefore only the link from IGFR to RAF was included and models were refitted excluding IKK inhibitor measurements for the further analysis to come.

3.3.2 Robustness assertion of model parameters

Having decided on a representative starting network structure, the reliability of the found parametrisation can be monitored. For this purpose 100 test sets were generated from the original data set by adding Gaussian noise. The model fitting procedure was run on all 100 sets for the starting network. Afterwards, the fully reduced network and the parameter distribution for the starting network was recorded. For intuitive visualisation identifiable parameter combinations were chosen by depicting all paths that start at a perturbed or measured node and end at a measured node (Fig. 3.9).

In both the pre- and post-reduction models the majority of parameter combinations exhibit a narrow distribution indicating an overall robust parametrisation by the algorithm. For the few exceptions that displayed a broad parameter variation the distribution was in most cases skewed to one side. If the median of the skewed cases was close to zero it hinted to a weak response and those paths were likely to be reduced by the algorithm, e.g. IRS-1 ending paths in HCT116 and ERK-GSK3A/B path in SW480. Additionally it could be observed that in many cases the less robust parameter of a particular cell line share a common node that has not been measured, as for example the EGFR node in HT29 cells or the PI3K node in LIM1215 cells. As these nodes fall into the non-identifiable combinations noise seems to affect their parametrisation more strongly.

In all five cell lines no pattern switch between pre- and post-reduction was apparent, i.e. wide distribution remained wide and narrow stayed narrow. Therefore the strategy to thoroughly sample the initial network and afterwards locally fit the reduced network with the parametrisation from the initial best fit seems to be robust. Furthermore the median values hardly changed between the pre- and post-reduction models implicating that the reduced models can be seen as the starting network model stripped of links that were unsupported by the data. The median values of the reduced model were also in good agreement with the parameter estimations of the model that was generated from the unnoised data (open left pointing arrows in the right diagrams), demonstrating that the original parametrisation is representative.

By investigating the overlap between removed links and the percentage of simulations resulting in non-zero parameters (as a measure of link confidence) a strong concordance could be observed. For the first four cell lines always the least confident links (lowest percentage of non-zero parameter values) were removed and in addition all removed links had been assigned a confidence of $\leq 50\%$. The only exception of this could be seen for SW480 cells where the more confident links ERK -> IRS-1 (78%) and AKT -> IRS-1 (59%) were removed but the least confident link p70S6K -> IRS-1 (31%) was kept. These three parameters are dependent on each other as ERK and AKT act directly on IRS-1 as well as indirectly over p70S6K. Under these circumstances the removal of the first link governs the following reduction procedure, which in the original network was the link between AKT and IRS-1. Hence, in this example keeping the hub was favoured over keeping direct links which is due to the greedy hill-climbing procedure. As all the derived coefficients are small the effect of this discrepancy might be of negligible size for the further analysis.

Overall it could be confirmed that the procedure robustly identified most parameters and retained those connections that bear the main information.

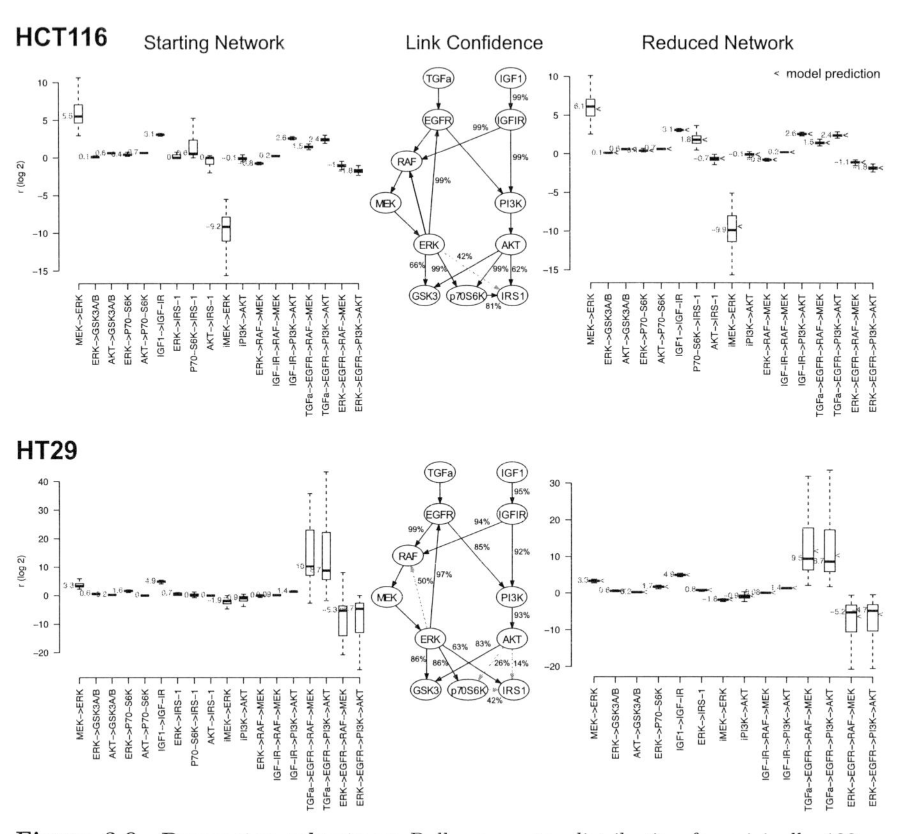

Figure 3.9 Parameter robustness Bulk parameter distribution for originally 100 data sets simulated from original measurements by adding random noise taking either the starting network structure (left) or the reduced structure (right). Outlier fits with χ^2 scores higher than 10 times the median χ^2 score were removed beforehand. Numbers indicate the median value and open arrows on the right boxplots indicate parameter estimations from the original model. The networks contain the percentage of non-zero links found in the starting network simulations if $\neq$ 100%. For clarity points outside the whiskers were omitted.

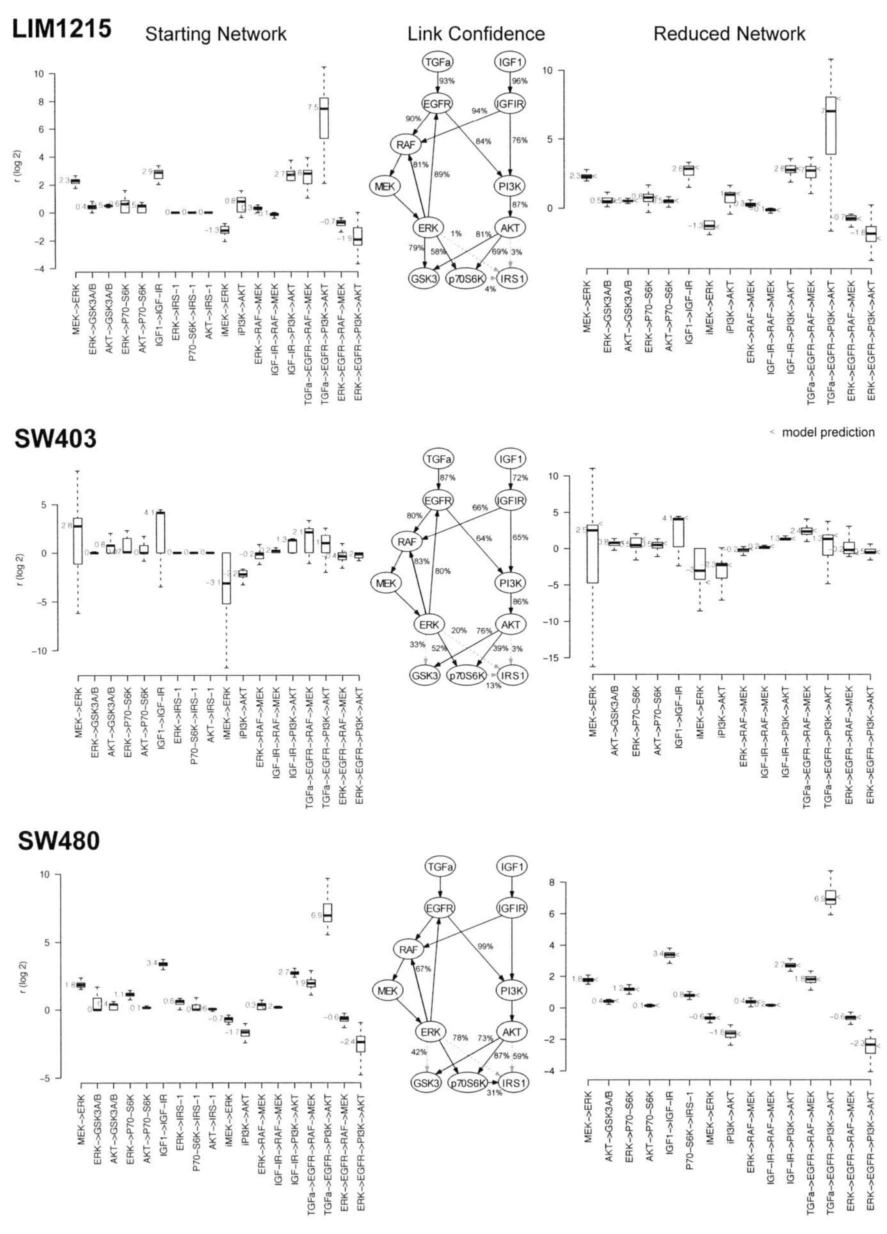

Figure 3.9 (continued) Parameter robustness.

3.3.3 Model fit uncovers differences in network structure

After the model fit has been evaluated from a technical perspective it can be investigated under biological aspects. Fig. 3.10A shows the experimental measurement side-by-side with the corresponding model fit. For each signalling node, measured phosphorylation is displayed in the upper row (indicated by filled triangle) and the model simulation in the lower row (empty triangle). Very good agreement can be seen for ligand stimulation and in general to nodes close to the receptors whereas in some cells few discrepancies can be observed for the PI3K inhibitor data and for downstream signals such as GSK3A/B. It can still be concluded that the resulting model could mimic most of the responses in quantitative detail.

The topology of the final models is displayed in Fig. 3.10B. Edges that the modelling procedure has removed in at least one cell line are depicted as dashed lines, with color-coded circles indicating the cell line. One interesting difference in the topology of the networks is the feedback from ERK to RAF, which has only been removed in HT29 cells. This is in line with previous findings that BRAF^{V600E} mutation disables MAPK feedback regulation to RAF (Friday et al., 2008; Sturm et al., 2010; Fritsche-Guenther et al., 2011) and HT29 is the only cell line harbour-

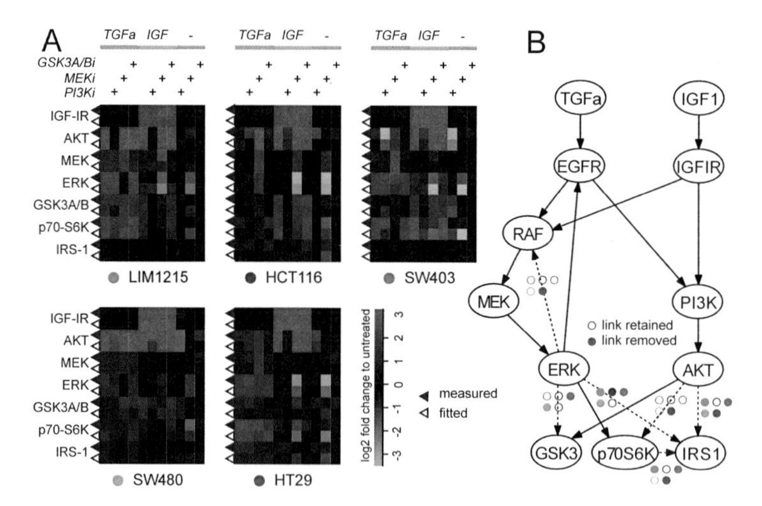

Figure 3.10 Model fit and differences in network topology (A) Heat maps for the five cell lines showing log2 fold changes of phosphorylation (filled triangles) and the corresponding model simulation (empty triangles). **(B)** Network structure derived from the model fit. Dashed lines indicate edges that are removed in those cell lines marked by filled circles (same colours and layout as in A).

ing that mutation (Table 3.1). Interestingly, the topology for nodes downstream of the output kinases ERK and AKT differs. For example, the phosphorylation site measured for IRS-1 is not connected to ERK and AKT in LIM1215, SW480 and SW403, and only connected to ERK in HT29 cells. Taken together, the model-fitting procedure allowed to identify qualitative differences in signal transduction networks in this cell line panel, of which one can be related to a specific mutation.

3.3.4 Quantitative differences between signalling networks

In addition to qualitative differences in the underlying network topology, it was also of interest to study how quantitative aspects of the signalling networks differ in these cells. Overall, the models contained on average 15 parameters ranging from 14 in SW403 to 17 in HCT116 cells. As these parameters typically correspond to combinations of response coefficients, the most intuitive way to inspect these parameter combinations was by recomputing these parameter combinations such that they correspond to paths in the network. The values of such parameter combinations are visualised in Fig. 3.11. Interestingly, many paths further downstream, such as the response coefficient from AKT to GSK3A/B or from ERK to p70S6K have comparable values in all cell line models. This suggests that cer-

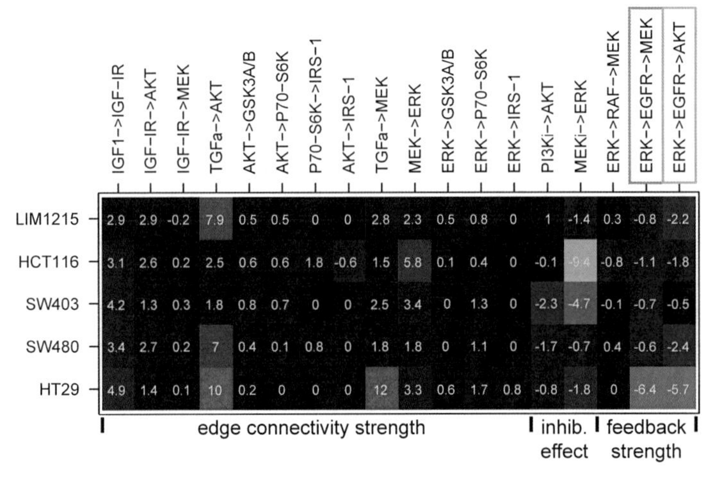

Figure 3.11 Cell-line specific signalling strength parametrisation Model-deduced parameter values representing signalling strength for identifiable paths, inhibitors and feedback/crosstalk expressed as log2 response coefficients. Orange and blue boxes indicate feedback from ERK via EGFR to MEK, and crosstalk from ERK via EGFR to AKT.

tain quantitative aspects of signalling are comparable in these cells, despite their heterogeneous genetic background. On the other hand, there are also strong differences between the cells, most of which relate to external perturbations. For example, TGFα does have a strong effect on MEK phosphorylation in HT29 cells, but can only weakly activate MEK in HCT116 cells. It might be worthwhile to test whether these observations stand in relation to receptor abundance or are due to the mutations within the pathway.

The ERK-RAF feedback is absent in BRAF-mutated HT29 cells and bears a negative sign in HCT116 and SW403 cells as described in the literature (Anderson et al., 1991; Dougherty et al., 2005). However, in the other two cell lines, SW480, and LIM1215, this feedback is predicted to exert a small positive effect. By looking at the other ERK-dependent feedback to EGFR (orange box in Figure 3.11) in those cells it can be seen that AKT is much more affected than MEK and therefore the algorithm may have tried to compensate for this discrepancy by assuming a positive RAF feedback. In fact in COS7 cells three ERK-dependent phosphorylation sites on c-RAF have been identified that increase CRAF activation (Balan et al., 2006). Thus the positive feedback assumption may be true, but to pinpoint this prediction, further measurements such as inhibition of the EGFR or RAF would be necessary. Regardless of the sign of the ERK-RAF feedback it can be perceived that on the observed time scale the effect of the upstream ERK-EGFR feedback is far more pronounced and seems superior to the RAF feedback.

All cells show negative response coefficients for feedback regulation from ERK via EGFR back to MEK (compare network in Fig. 3.10), with HT29 cells containing the strongest feedback. An interesting aspect of this ERK-EGFR feedback is that it connects ERK signalling with AKT activity in an EGFR-dependent manner, as ERK inhibits the EGF receptor that also stimulates AKT signalling. Thus, inhibition in the ERK pathway can lead to an activation of AKT if ligands for the EGFR receptors are present. This effect has been previously described as a mechanism of drug resistance in tumour cells with BRAF mutation (Prahallad et al., 2012). When inspecting the response coefficient of the path from ERK to AKT via EGFR, high negative figures for HT29, the BRAF-mutated cell line, could be found (blue box in Fig. 3.11). Intriguingly, the model fit unveiled that the feedback is present in all cell lines and causes strong crosstalk from ERK to AKT in all cell lines, including those harbouring KRAS mutations (blue box in Fig. 3.11; comp. Table 3.1). This finding prompted further investigations as the ERK-AKT crosstalk might be crucial in targeting KRAS-mutated tumours.

3.4 Model assisted study of the ERK-AKT crosstalk

If the ERK-AKT crosstalk is indeed as strong as predicted, downregulation of ERK would cause a paradoxical surge in AKT activation. Presuming that cells need both pathways for growth and survival it might be worthwhile to devise treatment combinations that break the cross-talk and prevent the hyperactivation of AKT. Therefore a strategy was followed that consisted of three steps:

1. Verify the predicted crosstalk and thoroughly investigate the properties.
2. Simulate suitable combinatorial treatments and test predictions on the signalling level.
3. Apply the most promising therapeutics to cell cultures or xenograft models and monitor growth.

3.4.1 Verification and characterisation of the crosstalk between ERK and AKT

To confirm the EGFR-dependent crosstalk between ERK and AKT, independent experiments were performed in HT29 and HCT116 cells, as examples for cells with BRAF and RAS mutations. Fig. 3.12A shows that AKT phosphorylation is only slightly increased when a MEK inhibitor is applied alone, and that AKT is only weakly stimulated by TGFα. In line with the crosstalk hypothesis, phosphorylation of AKT was significantly increased ($\alpha = 0.05$) by a factor of 2 - 3 when cells were pre-treated with the MEK inhibitor and stimulated with TGFα, confirming that the crosstalk operates irrespective of mutations in BRAF or RAS. Stimulation with EGF, another specific ligand of the EGF receptor, for 10 minutes, resulted in a similar observation (Figure 3.12A).

The question arose whether the negative feedback directly acts at the level of the EGF receptor, or more generally de-sensitises growth factor receptors, for example by acting on shared adaptor molecules (Dhillon et al., 2007). To disentangle the network that mediates the feedback, cells were stimulated with the growth factors HGF, FGF, and IGF which do not signal via EGFR (Fig. 3.12B). In contrast to EGF, pre-inhibiting either cell line with AZD6244 had no significant effect on the AKT response to HGF. Similarly, it had no effect on the response to FGF in HT29 cells, and only a weak effect in HCT116 cells. When using IGF as a control that primarily stimulates AKT, pre-treatment with MEK inhibitors did not change AKT phosphorylation level. This suggests that the feedback common to HCT116

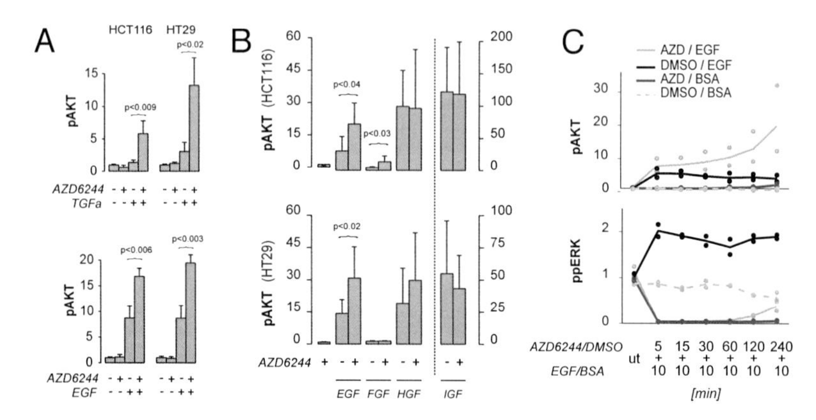

Figure 3.12 ERK-AKT crosstalk is mediated by EGFR and independent of RAS or RAF mutation (A) Increase of AKT phosphorylation in HCT116 (KRAS mutation) and HT29 (BRAF mutation) cells incubated with MEK inhibitor AZD6244 (0.1 mM) or its solvent control (DMSO) for 1 h before application of TGFα or BSA for 30 min. Effect can be also seen with 10 min EGF treatment in combination with AZD6244 (1 mM). **(B)** Increase of Akt phosphorylation compared with solvent control (DMSO) with combinatorial treatment of AZD6244 (5 mM) and different growth factors (10 min). **(C)** Response of phospho-ERK and AKT to 10 min EGF treatment for different pre-incubation times of AZD6244 (1 mM) in HCT116 cells. BSA and DMSO are solvent controls for EGF and AZD6244, respectively. All data measured using Luminex assays and shown as fold change. Brackets indicate significant one-sided t-test, $\alpha = 0.05$. Error bars indicate s.d. of n=3 samples. *Raw data generated by Anja Sieber and Raphaela Fritsche.*

and HT29 acts on the EGFR and not on downstream adaptors that are shared by many growth factor receptors.

It was next aimed to define the time scale on which the feedback operates. HCT116 cells were treated with the MEK inhibitor for various times and were then stimulated with EGF for 10 minutes and AKT and ERK phosphorylation was measured (Fig. 3.12C). Interestingly, already with a total inhibition time of 15 min, AKT phosphorylation was increased, which rose further for longer pre-inhibition times, suggesting that the feedback operates as an integral feedback. Notably, after about 2 h, also non-EGF-treated but MEK-inhibited cells showed a slight increase of AKT levels, indicating an EGFR-independent feedback operating on a longer time scale. Furthermore, the EGFR-dependent feedback shows that EGF treatment can partially restore the pre-inhibitor phospho-ERK level after $\approx$ 4h of inhibition. It can be speculated that this increase is achieved by the ERK-CRAF feedback that causes an accumulation of active MEK (Friday et al., 2008; Pratilas et al., 2009) to such an extend that the effect of the inhibitor is overwhelmed and

ERK gets reactivated. In fact, a reactivation of ERK after MEK inhibition has been observed for many MEK inhibitor-treated patients and a double treatment with MEK and RAF inhibitor was suggested to break this resistance (Poulikakos and Solit, 2011).

3.4.2 Prediction of potentially effective double inhibitions

The model was then used to predict combinatorial treatments that reduce ERK activity without activating AKT. As the initial model encompassed only two inhibitors directly positioned in the EGFR pathway, it was decided to re-train the model to include further inhibitors against EGFR (gefitinib) and RAF (sorafenib) for a KRAS (HCT116) and BRAF (HT29) mutated cell line (see Fig. 3.13A/B). During the re-training only the parameters representing the inhibitory strength of the four inhibitors were allowed to be altered while the other parameters were kept at the level of the original whole model fit. In order to obtain a better picture of the reliability of the results, not the original fit but the 100 noised simulations of the original data sets (see Figure 3.9) were re-fitted to the new data. Thus the variability of the fitted response (Fig. 3.13A) and the parameter range (Fig. 3.13B) could be estimated. The fitted response range was shown to be mostly narrow and the new data could be well adapted by only adjusting the inhibitor strengths even in combination with TGFa stimulation. The parameter range of the inhibitors showed that the strength of the MEK and RAF inhibitor could be defined very precisely in both cell lines giving good reasons to believe that those parameters were reliably and reproducibly estimated. PI3K and EGFR on the other side showed a much wider distribution for both cell lines albeit with different distributions. Possibly those variations could be prevented by re-fitting the identifiable parameter combinations and not just the single parameter that seemed to be less noise robust. To be able to re-fit the identifiable paths, however, a larger data set would have to be generated that was not available in this study.

The retrained model was now utilised to predict treatments of inhibitors of RAF or MEK combined with EGFR or PI3K plus the combination of the latter two (red bars in Fig. 3.13C). For all these combinatorial perturbations, the model predicted a reduction in AKT activity when compared with TGFα treatment only. As expected the phospho-ERK was strongly downregulated for all selected combinations which was captured nicely by the model predictions. Measurement of phospho-AKT confirmed most predictions with the exception of the two combinations of MEK/RAF inhibitors combined with PI3K inhibitor in HT29 cells (black

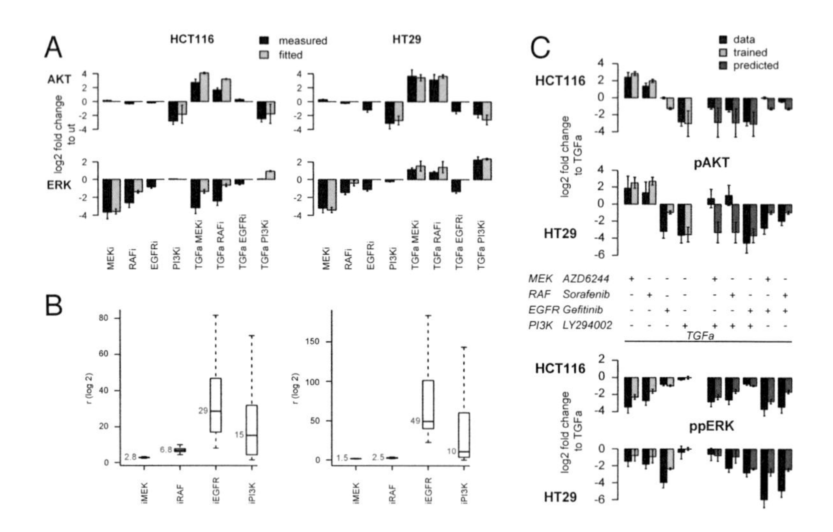

Figure 3.13 Model retraining for additional inhibitors and prediction of suitable inhibitor combinations (A) New data (black bars) was generated to determine the parameters for inhibitors against RAF, EGFR, MEK, and PI3K in the two indicated cell lines. RAF and EGFR inhibitors where newly introduced to the model and MEK and PI3K inhibitors were re-fitted (blue bars). Thereby only parameters for the inhibitions were allowed to change, retaining all other parameters of the original model. Data bars reflect the average of $n = 2$ with the error bars estimated by the error model (see B.1 on page 130). To obtain robust values from the original dataset (Fig. 3.1) 100 parameterisations were gained by adding Gaussian noise (cf. Figure 3.9) of which mean and standard deviation of the re-fits are shown. **(B)** Log2 parameter distribution for the indicated inhibitors. **(C)** Model prediction of selected double inhibitions (red bars) and Luminex measurements (dark bars) for phosphorylation of AKT and ERK in HCT116 and HT29 cells in the presence of TGFα in log2 scale. Model fits for single perturbations are shown on the left as blue bars. *Raw data measured by Anja Sieber*

bars in Fig. 3.13C). Since the ERK response was fitted correctly either experimental variation or other effects not included in the model caused this discrepancy.

Now it was sought for a suitable combinatorial treatment that reliably reduced the reactivation of AKT (Fig. 3.13C) and also blocked pre-stimulation-active ERK (Fig. 3.13A). From the prediction as well as in the measurements it appears that the combinations of MEK or RAF inhibitors together with EGFR inhibitor are as potent as when combining the former two with an PI3K inhibitor. The combination of EGFR and PI3K inhibition, although similarly strong in blocking AKT, can not target the basal ERK activation caused by the mutation and should therefore not result in a different response than the PI3K inhibition alone.

3.4.3 Combined inhibition of EGFR and ERK prevents growth in various tumour cells

It was now important to see, whether the newly discovered combination treatment of MAPK inactivation and EGFR inhibition also stops proliferation of tumour cells. To assess this, cell growth of HCT116 and HT29 cells was measured with those four inhibitors alone, their vehicle control DMSO, and in selected combinations. For this study a microelectronic real-time cell analyser system measuring the change of the electrical impedance of adherent cells growing on electrodes was used. Cell index, which is proportional to cell number (Witzel et al., 2015), was measured every 20 minutes starting one day before and ending five days after treatment.

Inhibition of the EGF receptor as well as inhibition of PI3K alone had no effect on proliferation in HT29 and HCT116 (Fig. 3.14A, top middle), and also combined application of both inhibitors did not alter proliferation either. This is in line with the notion that the oncogenes KRAS and BRAF drive proliferation in these cells via MAPK signalling and do not require the EGFR to generate the pro-proliferative signal. It was next investigated how manipulations downstream of their driver mutations alter proliferation in combination with PI3K inhibition. To

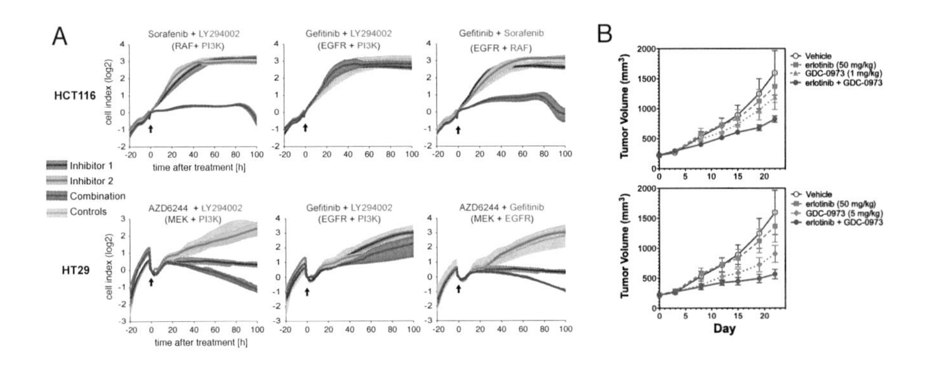

Figure 3.14 Model-derived combinatorial treatment options prove to be effective combinations *in vitro* and *in vivo* (A) Cell proliferation (xCELLigence) of HCT116 and HT29 cells in response to single and double inhibition and control. Arrows mark time of treatment, and areas represent the maximal range of the growth curves, with dark lines reflecting the mean ($n \geq 3$). **(B)** tumour growth of colorectal cancer cell line DLD-1 transplanted into nude mice, which received a daily oral gavage of one of two different concentrations of MEK inhibitor GDC-0973, EGFR inhibitor erlotinib alone or combined. Error bars represent s.e.m. with $n = 10$. *(A) raw data by Raphaela Fritsche, evaluation by Franziska Witzel (B) raw data and evaluation Leanne Berry*

inhibit the MAPK signalling pathway, it was decided to use two different inhibitors in the two cell lines. Based on the differences in ERK-RAF feedback (Sturm et al., 2010; Fritsche-Guenther et al., 2011), inhibition of MAPK activity with the RAF inhibitor sorafenib in HCT116 cells and the MEK inhibitor AZD6244 in HT29 cells seemed most appropriate. This decision is further supported by the corresponding measurements of the double inhibition in Fig. 3.13C, where those combinations showed the strongest effect.

Application of sorafenib alone had no effect on growth of the RAS/PI3K mutant cell line model HCT116, suggesting that growth may depend on multiple redundant pathways. However, in combination with a PI3K inhibitor, the cells stopped growing, indicating that MAPK and AKT signalling redundantly control cell growth (Fig. 3.14A, top left). As the model predicts that a combination of RAF and EGFR inhibitor prevents AKT activation, the EGFR inhibitor gefitinib was given together with sorafenib. In line with the model, this combinatorial treatment synergistically reduced growth as strong as the combination with PI3K inhibition (Fig. 3.14A, top right). In BRAF mutant HT29 cells, MEK inhibition blocked growth, reproducing previous observations (Solit et al., 2006), and PI3K inhibition alone had no effect, confirming that HT29 cells depend solely on MAPK signalling for growth. The combination of MEK and PI3K inhibitor led to a decrease of the cell index (Fig. 3.14A, bottom left). The model predicts that EGFR would also act synergistically with MEK inhibition to block AKT activation. In line with this prediction, EGFR inhibition had no effect when provided alone, but caused a stronger decrease in cell index when cells were treated in combination with a MEK inhibitor than any single component (Fig. 3.14A, bottom right).

To further test the effects of breaking the ERK-EGFR feedback in a more natural environment a xenograft model cell line DLD-1 that harbours both KRAS^{G13D} and PIK3C^{E545K} mutations was treated by collaborators at San Francisco. DLD-1 tumours were established in nude mice and treated with erlotinib (dosed at a clinically relevant dose of 50 mg/kg), GDC-0973 (a potent MEK-1/2 allosteric inhibitor, dosed both at 1 and 5 mg/kg), or the combination of erlotinib and GDC-0973 at both dose levels. Erlotinib and the lower dose of GDC-0973 (1 mg/kg) were not or only moderately effective as single agents resulting in 1 and 22% tumour growth inhibition (%TGI), respectively. The combination, however, demonstrated superior combination efficacy over either single agent alone with 36% TGI (Fig. 3.14B, top panel). The 5 mg/kg dose of GDC-0973 demonstrated better single-agent activity with 42% TGI, however, with 60% TGI combination with erlotinib demonstrated superior tumour inhibition to either drug used alone (Figure 3.14B, bottom panel).

The combination of these drugs was well tolerated in mice with minimal weight loss. Consequently a co-targeting therapy with EGFR inhibitor and low doses of MEK inhibitor yielded almost the same tumour reductive effect than a five-fold higher dose of the single MEK inhibitor treatment. This indicates that by suitable combinatorial treatments the medication burden and thus most likely the side-effects could be reduced without lowering the chances of therapeutic success.

Taken together, the results suggest that the EGF receptor is not required for maintaining the cellular phenotype in cells with RAS and RAF mutations neither *in vitro* nor *in vivo*. However, once MAPK signalling is blocked, the EGFR increases AKT activity and thus regains its key role in the network that was previously lost due to downstream mutations.

3.5 Discussion

A huge body of detailed mechanistic understanding about signalling processes has been accumulated within the last decades (Oda et al., 2005). Still it remains challenging to predict how a signalling network reacts when targeted therapies are applied due to inter-pathway cross talks and feedback loops (Kim et al., 2007; Friday et al., 2008; Sturm et al., 2010; Fritsche-Guenther et al., 2011; Kholodenko et al., 2012). The present study addressed this problem by quantifying MAPK/AKT signalling networks in a panel of colorectal cancer cell lines with differing genetic background. In here an experimental-computational pipeline was developed that compiles mathematical models for individual cell lines based on systematic perturbations. The complexity of the model was chosen such that it could be parametrised with a limited perturbation data set, but allowed to reveal network features such as feedback loops that simpler representations cannot account for. MRA generally requires that the response of the system to perturbations can be modelled using linear equations, and the system is close to steady state. Thus, the modelling procedure can be used to interpret perturbation screens, but parameters should be interpreted in a phenomenological rather than in a precise mechanistic way. Consequently, the procedure is helpful to interpret perturbation data and generate new hypotheses that can be subsequently tested. In this study, the majority of parameters could be well estimated from the data. However, if there are strong uncertainties in the parameters, methods such as MCMC (Hastings, 1970) or the profile likelihood (Raue et al., 2009) method can be readily applied to model parameters or structural uncertainties.

By comparing the signalling maps between the cell lines, it could be observed that the core signalling network was quantitatively very similar in all cells, although these cells were heterogeneous in their genetic constellation. This suggests that it is instrumental to build generic models of signalling despite genetic heterogeneity. Nevertheless, certain quantitative aspects differed strongly between cells, such as the strength of the response towards stimulation of the EGFR. The method discovered also qualitative differences between cell lines, such as the loss of the ERK-RAF feedback in HT29 cells, which can be traced back to the BRAF V600E mutation (Friday et al., 2008). Similar studies on larger cell line collectives may unveil further differences in network wiring due to the underlying mutations. Despite the diversity of mutations in the EGFR signalling network, a strong feedback from ERK to EGFR was found to be conserved in all five cell lines. At least in HCT116 cells, this feedback was detectable within 15 minutes but its strength increased further on time scales of hours, suggesting that it operates as an integral feedback, which may play a role in signalling homoeostasis (Yi et al., 2000; Prahallad et al., 2012). Different mechanisms of feedback regulation of the EGFR are known, which may all contribute. Transcriptional feedbacks such as mediated by MIG-6 (Yoon et al., 2012) or the Sprouty family (Mason et al., 2006) are acting too slow for the fast feedback regulation, but may be involved in later phases. Posttranslational regulations of adaptors shared between receptors, such as SOS (Douville and Downward, 1997; Shankaran and Wiley, 2010) or Gab1 (Yu et al., 2002), are fast enough but most likely not the main feedback players, as the cross talk was only strong when EGFR was stimulated, but was not or only weakly present when HGF, FGF or IGF were given. Thus, the strongest target of this feedback is most likely EGFR itself. It has been reported from pancreatic cancer cells that EGFR exhibited increased phosphorylation at Y1068, Y1045 and Y845 when MEK was inhibited (Gan et al., 2010). Another study reported that phosphorylation of T669 by ERK affects EGFR turnover (Birtwistle et al., 2007). Knockdown of phosphatase CDC25C did also lead to increased phosphorylation of EGFR at Y1068 and it was suggested that ERK changes CDC25 activity and by this regulates phosphorylation of EGFR (Prahallad et al., 2012).

A consequence of this feedback for targeted inhibition is that it leads to activation of AKT upon inhibition within MAPK signalling. Such feedback-mediated crosstalk has been noted in many different tumours such as breast cancer (Mirzoeva et al., 2009; Lu et al., 2011), prostate cancer (Gan et al., 2010), melanoma (Gopal et al., 2010), gastric cancer, (Yoon et al., 2009) and colorectal cancer (Prahallad et al., 2012). Increased phosphorylation of AKT may cause drug resistance, as

AKT activity stimulates survival and migration. Furthermore, AKT shares a complex network of transcription factors with ERK (Stelniec-Klotz et al., 2012) and both paths converge on key proteins important for cellular function such as cyclin-D1 for growth control (Halilovic et al., 2010) and Bad for apoptotic regulation (She et al., 2005), explaining the need to switch off both pathways to efficiently target tumour cells.

The MRA model allowed devising combinatorial therapies that block ERK activation and at the same time prevent rise of AKT activity. Simulations showed that while PI3K inhibitors may be used to block AKT signalling, EGFR inhibition could also prevent strong AKT hyper-activation, irrespective of whether RAF, RAS or PI3K was mutated which was then verified experimentally. In line with this prediction, approximately the same synergistic reduction of growth for PI3K/MAPK combination could be observed as for EGFR/MAPK inhibition *in vitro* in two cell line models. Therefore, while upregulation of AKT by MAPK inhibition can be successfully blocked by PI3K or mTOR inhibition (Balmanno et al., 2009; Mirzoeva et al., 2009; Aksamitiene et al., 2010) in colorectal cancer models, upstream inhibition of EGFR may be similarly potent with possibly less side effects. Using a RAS/PI3K mutant colon cancer xenograft model, this study could confirm that combination of inhibitors against MEK and EGFR is also an efficient therapy *in vivo*. This extends previous findings showing that BRAF inhibitors synergise with EGFR inhibition in a colorectal cancer xenograft model with BRAF mutation (Prahallad et al., 2012).

For RAS-mutated tumours, so far no targeted therapy is available in the clinics (Baines et al., 2011; Ward et al., 2012), and a mutation in RAS was generally assumed to preclude EGFR-directed interventions. Our results suggest that RAS-mutated tumour cells can be successfully treated by EGFR inhibitors if provided together with MEK or RAF inhibitors. While our model can predict successful combinatorial treatments, it has limitations. It cannot account for combinations that require sequential application of drugs (Lee et al., 2012), and fails to capture resistance due to tumour-stroma interactions (Sebens and Schafer, 2012). It is likely that in other cell types different combinations of drugs may be more successful, as the role of specific feedbacks can vary in different cell types, and can even switch between positive and negative effects depending on receptor expression levels (Birtwistle et al., 2007).

Tumour evolution is one of the major causes for eventual relapse (Iwasa et al., 2006). For example, relapse after long-term treatment with the anti-EGFR agents panitumumab in KRAS wild-type colorectal carcinoma (Amado et al., 2008; Karapetis et al., 2008) is often caused by selection for mutations downstream of EGFR (Diaz et al., 2012; Misale et al., 2012). The combinatorial treatment predicted by the model presented in here may thus be advisable even for EGFR-addicted tumours, as it will counteract selection for additional mutations in RAS or RAF. This proposed treatment of MEK inhibition with EGFR inhibition to overcome EGFR resistance has indeed been proven in an extensive drug screen effort (Misale et al., 2014), demonstrating the universality of the modelling results. In colon cells, the EGFR receptor is very potent, but depending on tissue and mutational status, other receptors may be more important in stimulating the network. For example, c-Met, that is as well feedback-regulated and signals to both ERK and AKT, was shown to mediate resistance to BRAF inhibition in melanoma (Wilson et al., 2012), suggesting c-Met as the most effective co-target in this particular situation. Hence, these results further highlight that downstream mutations such as in RAS and BRAF do not necessarily invalidate upstream drug treatment if provided in combination with a suitable downstream inhibitor.

3.6 Concluding Remarks

So far no reverse engineering approach has emerged as the ultimate tool to derive network characteristics and kinetics from perturbation data. In the literature, it is repeatedly emphasised that MRA is in principle capable of modelling signalling networks (Kholodenko et al., 1997; Kholodenko, 2000; Kholodenko et al., 2002; Bruggeman et al., 2002; Andrec et al., 2005). Next to numerous theoretical models, one study succeeded in reverse engineering the three-tired MAPK cascade RAF/MEK/ERK using real experimental data (Santos et al., 2007). In Santos et al. (2007) systematic perturbations were achieved by whole isoform knockdown and then the coefficients were estimated after short-term stimulations by two different ligands, EGF and NGF. This concept was taken one step further in this chapter by estimating a larger network and by deriving response coefficients for inhibitory, stimulatory and combinatorial responses from incomplete perturbation data. Furthermore, next to simulating and analysing the parameters of the model, in here also consequences of novel inhibitor combinations could be modelled and were in good agreement with the subsequent measurements. The key finding heavily relied on the fact that MRA is able to integrate the effect of feedbacks which results in more realistic predictions. In principle this approach can be applied to any perturbation data set, however one has to keep in mind that due to the steady state assumption resulting parameters can vary in time and instead of representing a kinetic constant they rather reflect a snapshot of the current signalling flow.

Another novel aspect of this study was that it was conducted on a heterogeneous cell line panel, which allowed to compare respective parameters and to detect differences in signalling as well as common paths which lead to the discovery of the importance of the EGFR feedback. Apart from the BRAF mutation in HT29, the cell lines could not be distinguished according to their inherent mutations. This can be due to the limited view on signalling by only perturbing at six positions and measuring eight signals. In addition, due to the study layout in the present study, it was impossible to compare the basic signalling levels of the cell lines to derive the mutational impact on the unperturbed state. However even with a larger data set with comparable basic levels at hand one will encounter problems in identifying the mutation mix when the cell lines are not isogenic but derived from different patients and - even worse - different cell-tissue origins. The best study case for this particular question would be to first monitor a set of isogenic cell lines differing only in one defined mutation comparing both the basic expression and the perturbed measurements. Under these conditions the difference in mutation is known and thus the perturbation that is present can be easily incorporated into the model as an additional node. If from those measurements a means has been found to reliably predict those mutations, the knowledge can then be applied to uncover driver mutations even in a heterogeneous cell line panel. The strategy to detect the mutation mix could then follow in a similar regime as the greedy hill model selection applied in here to alter the network structure.

The biological consequences of the findings in this study are manifold. Heterogeneous cells still employ common pathways which can be targeted to treat a wider range of cancer cells. The thusly found treatment might be less prone to emergence of resistant mutations. Although the precise molecular mechanism that mediates the feedback could not be disclosed, the EGFR could be encircled as the most likely involved receptor. As an alternative mechanism to the putative change in the phosphorylation pattern the internalisation or recycling speed of the EGFR might also be changed (Schoeberl et al., 2002; Wiley, 2003). Follow-up studies in our laboratory, however, could not detect an altered internalisation behaviour (unpublished work). One way to close the knowledge gap between ERK inactivation and EGFR re-sensitisation is the detection and quantification of EGFR-bound proteins before and after ERK inhibition via methods such as co-immunoprecipitation and subsequent mass spectrometry. There is good reason to believe that negative feedbacks leading to cross-activation exist for many other receptors and adaptor proteins such as IRS-1, which has been described to mediate a crosstalk between mTOR/PI3K and JAK/STAT (Britschgi et al., 2012). Other studies in breast cancer cells identified a long-term AKT-receptor tyrosine kinase expression feedback via FOXO3 (Chandarlapaty et al., 2011) on a scale of 24 h after inhibition that could also activate ERK (Serra et al., 2011).

In the current study a systematic experimental setup was introduced that in combination with the modified MRA represents a potent tool to detect and quantify more crosstalks in signalling. Identifying those structures in future works will be crucial to get closer to the goal of building a valid virtual cell model and to better predict the outcome of single and combinatorial treatments. The here proposed combinatorial treatment is a promising therapeutic option in BRAF and KRAS-mutated colon cancer cells with the latter still lacking an effective therapy. Recent studies showed that the scientific community is moving closer to develop a means to block mutated RAS (cf. Cox et al. (2014)) with either small molecule inhibitors (Zimmermann et al., 2013; Ostrem et al., 2013) or by siRNA capsules located near the damaged tumourigenic tissue (Khvalevsky et al., 2013). It remains to be seen whether these treatments will be robust against the EGFR feedback in heterozygous mutation bearing cells as the non-mutated RAS can still relay the EGFR signal as well as the other isoforms of RAS. Additionally, single agent RAS treatment remains less robust to emerging mutations downstream (e.g. at BRAF or PI3K) or in case the signal can still be relayed it will be even vulnerable to upstream mutations at the receptor or ligand level. Therefore the search for suitable combinatorial treatments seems to be the most robust strategy to win the fight against cancer progression.

Conceptual modelling resolves role of TTP in HIF-1 regulation

This work has partly been published in Fähling et al. (2012). Contributions of co-authors are indicated. For details on experimental methods refer to the materials and methods section of the original article.

Synopsis

Due to the property of MRA to obtain instantaneous predictions from complex network structures, its use in model-supported hypothesis testing was evaluated. In this chapter MRA is used to investigate the role of the RNA binding protein Tristetraprolin (TTP) in normoxic regulation of Hypoxia inducible factor 1 (HIF-1). In a microarray survey it was discovered that TTP mRNA levels showed strong positive correlation to HIF-1 target mRNAs, in spite of its demonstrated molecular role as a destabiliser of HIF-1α mRNA. In order to explain this paradox observation a set of hypotheses was generated and simulated using MRA as modelling framework. The model-assisted approach helped to identify a dual role of TTP in the context of HIF-1 regulation which is controlled by phosphorylation of TTP.

4.1 Introduction

In the previous chapter an ideal combination of experiments and mechanistic modelling was shown. First, the model type was chosen, then the methodology developed and only afterwards the experiments were carried out in concordance with the selected modelling scheme. However, in most cooperations between experimentalists and theoreticians it is usually the other way round. First, an experimental observation is made that either appears to be counterintuitive to previous findings or lacks explication due to insufficient knowledge. In search of an an explanation, a set of suitable models is generated whose predictions can then be verified or falsified by follow-up experiments. The modelling challenge in this so-called bottom-up approach resides in the fact that data and information is scarce. In this case coarse grain models can be of use as the models often have to explain phenomenological observations.

In the following, MRA will be utilised as a conceptual modelling tool to disentangle the regulatory layers on which the RNA binding protein Tristetraprolin (TTP) can regulate activity of HIF-1. In order to properly introduce the biological background of the modelling subject, regulation and function of HIF-1 will be briefly discussed. This will be followed by the delineation of the biological puzzle that was then sought to be solved by MRA models.

4.1.1 Molecular function of HIF-1

Hypoxia inducible factor 1 (HIF-1) is a transcription factor that plays a central role in controlling glucose metabolism and angiogenesis and further participates in cell proliferation and survival (reviewed in Semenza (2003)). HIF-1 is a complex comprised of the two subunits HIF-1α and HIF-1β. The latter subunit is found to be constitutively expressed and stable, whereas the former encodes a largely unstable protein. Thus, activity of the HIF-1 complex is predominantly dependent on the availability of HIF-1α.

The best investigated role of HIF-1 is the coordination of cellular response to changes in oxygen level. During normoxy, HIF-1α protein is rapidly degraded exhibiting a very short half-life of 5-8 min (Qutub and Popel, 2006). This fast degradation is triggered by O_2-dependent hydroxylation, which then leads to subsequent ubiquitination. When the oxygen level is declining, non-hydroxylated HIF-1α accumulates allowing it to form transcriptionally active dimers with HIF-1β. Active HIF-1 upregulates glycolytic enzymes to increase O_2-independent glycolysis and at the same time induces genes such as Pyruvate Dehydrogenase Kinase 1 (PDK1) to decrease aerobic respiration. In addition, HIF-1 activates genes such as Vascular Endothelial Growth Factor (VEGF) that induce blood vessel neogenesis (reviewed in Pouysségur et al. (2006)). In this fashion HIF-1 enables the cell to cope with conditions of low oxygen and at the same time triggers processes that alleviate hypoxic conditions[1].

The function of HIF-1 is often found to be abused by cancer cells (Cairns et al., 2011). Many tumours exhibit an up to 200-fold higher glycolysis rate than normal tissue which results in an increased ATP production rate of $\approx$ 10-13% (Koppenol et al., 2011). The abnormal cellular fermentation in the presence of oxygen is known

[1]Also other processes manage the hypoxic response, e.g. selective translation by mRNA partitioning (Staudacher et al., 2015)

as the Warburg effect (Warburg, 1925, 1956) and has been ascribed to HIF-1 hyperactivity (Kim and Dang, 2006). Next to the metabolic switch the angiogenesis potential of HIF-1 has been exploited as well by many cancer cells to boost tumour size (Harris, 2002). Consequently, abnormal upregulation of HIF-1 activity is commonly accepted as one of the hallmarks of cancer (Hanahan and Weinberg, 2011; Semenza, 2010b). Some instances for HIF-1 hyperactivation could be related to loss-of-function mutations of negative translational or posttranslational regulators such as the E3 ubiquitin ligase VHL (see Semenza (2010a)). However, as these mutations only occur in few cancer types in higher frequencies other means of HIF-1 deregulation are likely to exist. Accordingly, this study sought to detect the potential of pretranslational regulators to alter HIF-1 activity.

4.1.2 Study of normoxic HIF-1 regulation reveals controversial observation

Whereas in hypoxia posttranslational regulators are acknowledged as the main activators of HIF-1, only little is known about processes that act at normoxic conditions (Kuschel et al., 2012). Normoxic induction has been observed by enforced protein synthesis of HIF-1α upon activation of PI3K and RAS pathways (Semenza, 2003). Transcription of HIF-1α was seen to be activated by lipopolysaccharide to allow the initiation of inflammatory responses before tissues become hypoxic (Eltzschig and Carmeliet, 2011). In this spirit, Fähling et al. (2012) embarked to elucidate the contribution of pretranslational effects on HIF-1 activity in normoxy.

It was first explored whether pretranslational regulation of HIF-1α had a significant effect on HIF-activity. Hence, a list of known HIF-1 target genes (defined in Semenza (2003)) was correlated with HIF-1α expression in a set of 1202 microarray experiments stored in the Stanford Microarray Database (Demeter et al., 2007). Indeed, in a large group of experiments a significant positive correlation could be observed and therefore pretranslational regulators must act onHIF-1. Consequently a search for possible regulation mechanisms that can alter HIF-1α mRNA was initiated.

A major part of RNA regulation can be attributed to RNA-binding proteins which often bind to regulatory sites in the untranslated regions (UTR) (Guhaniyogi and Brewer, 2001; Wu and Brewer, 2012). When closer inspecting the 3'UTR of HIF-1α mRNA, Fähling et al. (2012) found that the UTR ranks among the 20 most conserved UTRs of 7953 genes tested. By deducing a functional role of the 3'UTR, its role in stability regulation was examined. The sequence of HIF-1α

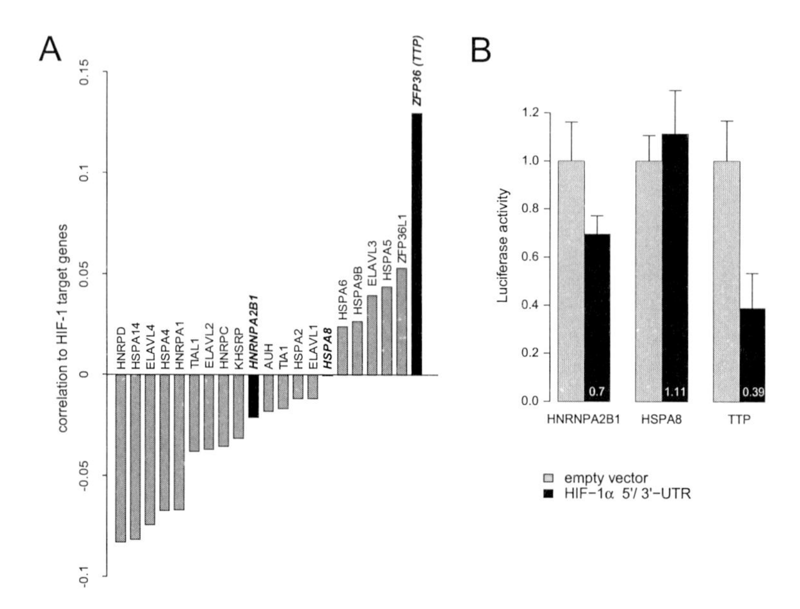

Figure 4.1 Controversial observations on TTP-HIF relationship (A) Correlation of gene expression of AU-rich element-binding proteins (ARE-BP) withHIF-1 target genes in a set of >1200 microarray experiments taken from Stanford Microarray Database. **(B)** HIF-1α mRNA UTR-dependent luciferase assay for three selected ARE-BPs co-transfected into HEK293 cells. The three ARE-BPs HNRNPA2B1, HSPA8 and TTP were chosen as representatives of groups with positive-, no- and negative correlation, respectively (marked in bold in A). The impact on HIF-1α UTR is in agreement with the observed correlation for all ARE-BPs, except for TTP.
(A) was generated by Ralf Mrowka and (B) was generated by Michael Fähling

mRNA 3'UTR includes a considerably high number of AU-rich elements (ARE) (Rossignol et al., 2002). These sequence motifs are known to be bound by ARE-binding proteins (ARE-BPs), prompting the decision to investigate the involvement of ARE-BPs in the regulation of HIF-1.

When correlating gene expression of known ARE-BPs with HIF-targets on the set of 1202 microarray experiments, the ARE-BPs could be divided into three groups exhibiting negative-, no- and positive correlation (Fig. 4.1A). Since correlation does not give information about directionality or even causality a subset of ARE-BPs chosen to reflect the three types of correlation was further tested for direct impact on HIF-1α. The influence on mRNA stability was monitored by co-transfecting a HIF-1α mRNA 5'/3'-UTR-fused reporter gene construct with representatives of either correlation group (Fig. 4.1B). In agreement with the observed correlation, negatively correlated HNRNPA2B1 conferred a destabilising effect and uncorrelated HSPA8 had no significant effect on HIF-1α UTR. In contrast to this the mRNA of TTP (ZFP36), that shows by far the most positive correlation of all ARE-BPs tested, is destabilising HIF-1α UTR to an even greater extend than the negatively correlated HNRNPA2B1. Accordingly, in the available literature TTP is described as a negative regulator of mRNA stability (Brooks et al., 2002; Blackshear, 2002; Stoecklin and Anderson, 2007) and destabilisation of HIF-1α mRNA by TTP has been confirmed by two independent studies (Kim et al., 2010; Chamboredon et al., 2011).

Therefore the question arose how TTP mRNA can be positively correlated with HIF-1 activity when its molecular function resides in the destabilisation of HIF-1α mRNA. In order to find a solution to this puzzle it was decided to develop and evaluate a set of literature-backed mathematical models.

4.2 Model-supported hypothesis testing

Model-assisted hypothesis testing is often applied to systems whose complexity or size defies human intuition. The modus operandi usually consists of four steps: (i) identify relevant actors and modes of actions, (ii) generate and simulate a set of biologically plausible minimal models, (iii) select the best explaining model(s) and (iv) devise feasible experiments to verify or falsify central properties. An important selection criterion for the best models is that they should use minimal additional information with at best biologically justifiable alterations. This constraint facilitates the testability of the models and further reduces the costs of model exclusion experiments.

In this study, the relevant actors are directly connected and consist only of the three central nodes TTP, HIF-1α and HIF-targets. The modelling objective is to devise models that produce positive correlation between TTP mRNA (ZFP36) and HIF-target mRNAs. As the time point and conditions in the set of 1202 microarray experiments is not confined, the most sensible approach is to study the correlation in steady state. Previously successful attempts for such a small scale model were accomplished by setting up a system of ODEs and sample unknown parameters, e.g. Cedersund et al. (2008); Blüthgen et al. (2009); Xu et al. (2010). The estimation of steady states from such ODE systems requires iterative runs until output convergence. Since this procedure has to be repeated for each initialisation it may be computationally intensive if many parameters are unknown, the network size is large or the convergence is slow. Conveniently, the global response matrix

R in the MRA context is defined as the steady state solution of a system of linear ODEs and can be derived by a single inversion of **r**. Therefore, for studies interested in steady state behaviour, the MRA framework is able to provide a ready-to-use format for hypothesis exploration.

4.2.1 MRA application scheme for hypothesis testing

The original concept of MRA states that the local connectivity **r** can be derived by systematically perturbing each node and measuring its global changes stored in **R** (Kholodenko et al., 2002; Bruggeman et al., 2002).

$$\mathbf{r} = -\Delta\mathbf{p}\mathbf{R}^{-1} \tag{4.1}$$

In here as well as in all other chapters this concept is reverted by trying to find a local response matrix that can best reproduce the global response matrix.

$$\mathbf{R} = -\mathbf{r}^{-1}\Delta\mathbf{p} \tag{4.2}$$

In this particular situation the aim is to find a local response matrix structure that is capable of producing global responses with positive correlation at the respective nodes. As the available data is scarce and mostly correlative, the possible network structures can only be tested qualitatively.

The observation to be described stems from a pool of 1202 microarray experiments. Therefore the actual local response coefficients are unknown and likely to be variable across all experiments. In contrast to this, the general regulation type, negative or positive influence, is not expected to change and can at most be reduced to zero, i.e. be absent. Therefore the network connectivity and sign information was used as a backbone to populate the local response matrix for each model scenario. Then, a Monte Carlo sampling was applied, which randomly generated instances of local response matrices that are compatible with the tested network. More specifically, the sign-conserved adjacency matrix was multiplied with a random number drawn from the log normal distribution $(0, \infty)$. Next the generated **r** and an arbitrary perturbation strength p_i were applied to Equation 4.2, yielding the steady state matrix **R**, from which the entries corresponding to stimulus-to-TTP mRNA (ZFP36) and stimulus-to-HIF-1 target mRNA were extracted (see models in Suppl. Sec. C 1 on page 133) and converted into binaries (1 for up and 0 for down). This procedure was repeated 10^6 times, collecting the binary responses for both mRNAs into vectors. Finally, as a measure of correlation, the Matthews correlation coefficient (mcc) was calculated from the two binary vectors (Baldi et al., 2000).

4.2.2 Model scenarios

Modelling the network around HIF-1 has been a subject for numerous studies (Kohn et al., 2004; Qutub and Popel, 2006; Dayan et al., 2009; Heiner and Sriram, 2010; Schmierer et al., 2010). However, all listed models were developed to simulate the response to variation of oxygen level or posttranslational regulators (Cavadas et al., 2013). In contrast to this, the majority of the correlated microarray experiments were not conducted at hypoxic conditions and even the negative effect of TTP on HIF-1α UTR shown in Fig. 4.1 was observed at normoxic conditions. Furthermore, effects of HIF-1α regulation other than posttranslational were not considered in pre-existing models. Thus, to investigate pretranslational regulations, models for the normoxic regulation of HIF-1α were devised independently of previous modelling attempts (Figures 4.2 - 4.5).

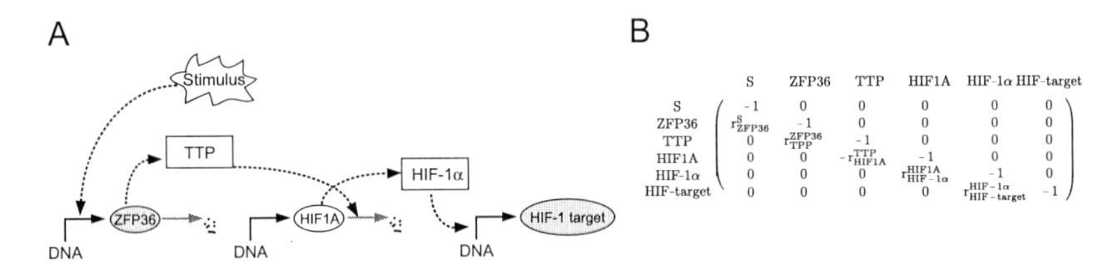

Figure 4.2 Basic model and realisation in MRA. (A) Graphical model of HIF-1 activity regulated by HIF1A stability. The system is driven by a stimulus S triggering transcription of TTP mRNA (ZFP36). Its protein product TTP destabilises the mRNA of HIF-1α (HIF1A). **(B)** Signed local response matrix of the model. Due to the linear action chain the entries of the global response matrix represent simply the product along the paths. Because of the negative local response coefficient between TTP and HIF1A, the global responses of the nodes, that are highlighted in grey in A, will always be negatively correlated, which cannot explain the results in Fig. 4.1A.

Figure 4.2 reflects the original perception that follows the notion that TTP acts as a purely negative regulator of HIF-1α. The model includes the production terms of HIF-1α and TTP protein from their mRNAs HIF1A and ZFP36, respectively. Furthermore the destabilisation of HIF1A by TTP is indicated to act on the degradation arrow in **A** and as local response coefficient $-\mathrm{r}_{\mathrm{HIF1A}}^{\mathrm{TPP}}$ with negative sign in **B**. To concentrate on the effect of HIF1A stability regulation on HIF-1 target gene expression, all other known regulations (such as translational- and posttranslational effects or complex formation) were assumed to stay constant in steady state. Transcriptional regulation of ZFP36 has been observed for many different stimuli such as SMAD (Ogawa et al., 2003), ELK1/EGR1 (Florkowska et al., 2012), and various inflammatory stimuli (Brahma et al., 2012; Chen et al., 2013). Thus the system was set up to be driven by a stimulus S that controlled ZFP36 transcription. The corresponding local response matrix in Fig. 4.2B shows a linear action chain and global responses can be calculated by multiplying along the path. Thus, for the depicted scenario, the global response of ZFP36 and HIF-1 targets to the stimulus S will be always anticorrelated due to the negative local response coefficient from TTP to HIF1A. In order to produce positive correlations three nested models were developed by extending the basic model with increasing complexity (see a complete delineation of the models in Suppl. Sec. C 1 on page 133).

(A) Negative feedback model TTP is described not just to bind and destabilise HIF1A but to exert the same effect on its own mRNA (Brooks et al., 2004; Tchen et al., 2004). Therefore a self-feedback was included in the model as a negative local response coefficient from TTP to ZFP36 (Fig. 4.3). When correlating the stimulus-driven global response coefficients of ZFP36 and HIF-1 targets (marked in grey), a strictly negative correlation could be observed for all parameters (Fig. 4.3B). By looking at the symbolic solutions of the two global response coefficients this becomes immediately clear as the negative feedback only reduces the overall strength

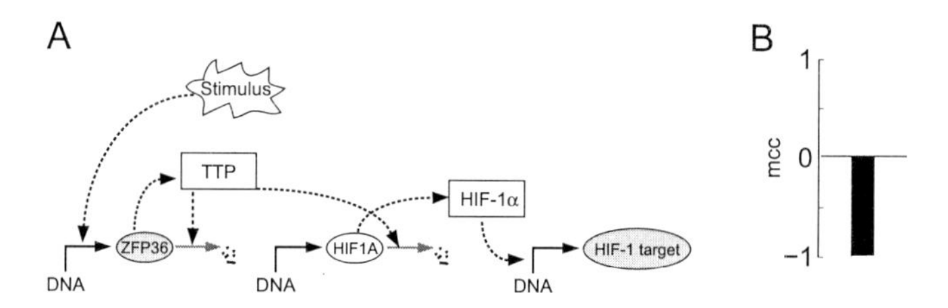

Figure 4.3 Negative feedback model. (A) Extension of Figure 4.2 by adding the finding that TTP destabilises not only HIF1A but also its own mRNA (ZFP36). **(B)** Model simulations predict that TTP mRNA is anticorrelated with HIF-1 target mRNAs (marked as grey ovals in A). Model parameters were chosen randomly in a Monte Carlo approach ($n = 10^6$). The frequency of positive and negative co-regulation of ZFP36 and HIF-1 targets was recorded and evaluated using the Matthews correlation coefficient (mcc).

but does not affect the sign of the respective R_i^js.

$$R^S_{ZFP36} = \frac{\mathrm{r}^{\mathrm{S}}_{\mathrm{ZFP36}}}{\mathrm{r}^{\mathrm{TTP}}_{\mathrm{ZFP36}}\,\mathrm{r}^{\mathrm{ZFP36}}_{\mathrm{TPP}} + 1} \quad , \quad R^S_{HIFtarget} = \frac{-\mathrm{r}^{\mathrm{S}}_{\mathrm{ZFP36}}\,\mathrm{r}^{\mathrm{ZFP36}}_{\mathrm{TPP}}\,\mathrm{r}^{\mathrm{TTP}}_{\mathrm{HIF1A}}\,\mathrm{r}^{\mathrm{HIF1A}}_{\mathrm{HIF1\alpha}}\,\mathrm{r}^{\mathrm{HIF1\alpha}}_{\mathrm{HIFtarget}}}{\mathrm{r}^{\mathrm{TTP}}_{\mathrm{ZFP36}}\,\mathrm{r}^{\mathrm{ZFP36}}_{\mathrm{TPP}} + 1} \tag{4.3}$$

Therefore the original conception that a negative regulator should be anticorrelated to the activity of its target, holds even in the presence of a negative feedback loop from TTP to ZFP36.

(B) TTP sequestration model Next, it was tested whether posttranslational modification of TTP changes the correlation. TTP is known to be phosphorylated at several sites (Carballo et al., 2001; Cao et al., 2004, 2006) which can be achieved by activating diverse kinases such as MK2, PI3K or JNK (Sandler and Stoecklin, 2008). This phosphorylation was further reported to reduce affinity of TTP to bind to AREs (Carballo et al., 2001; Hitti et al., 2006). Therefore the phosphorylation was included in the model by including an extra node pTTP, representing the phosphorylated form of TTP (Fig. 4.4A). The phosphorylation was realised by introducing a response coefficient $\mathrm{r}^{\mathrm{S}}_{\mathrm{pTTP}}$ that represents the phosphorylation and a response coefficient $\mathrm{r}^{\mathrm{S}}_{\mathrm{TTP}}$ with equal strength but opposite sign that accounts for the loss of unphosphorylated TTP. In this manner TTP is transformed into inactive pTTP and therefore sequestered from target mRNAs.

Model simulations showed that this effect on average decouples the correlation between ZFP36 and HIF-targets gene expression (Fig. 4.4B). This indicates that there are equal number of parameters with positive or negative correlation. The requirements for positive correlation can be deduced from the symbolic represen-

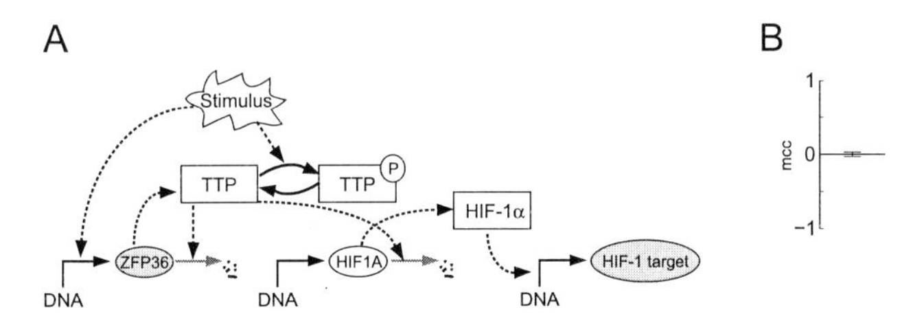

Figure 4.4 TTP sequestration model. (A) Extension of Figure 4.3 by incorporating phosphorylation of TTP. Phosphorylated TTP is assumed to lose binding affinity to 3' UTR and therefore TTP destabilisation is weakened. **(B)** Model simulations predict that the mRNA levels of TTP and HIF-1 targets become uncorrelated.

tations of the global response coefficients.

$$
\begin{aligned}
R^S_{ZFP36} &= \frac{\mathrm{r}^{\mathrm{S}}_{\mathrm{ZFP36}} + \mathrm{r}^{\mathrm{S}}_{\mathrm{TTP}}\,\mathrm{r}^{\mathrm{TTP}}_{\mathrm{ZFP36}}}{\mathrm{r}^{\mathrm{TTP}}_{\mathrm{ZFP36}}\,\mathrm{r}^{\mathrm{ZFP36}}_{\mathrm{TPP}} + 1}\;, \\
R^S_{HIFtarget} &= \frac{\mathrm{r}^{\mathrm{HIF1A}}_{\mathrm{HIF1}\alpha}\,\mathrm{r}^{\mathrm{HIF1}\alpha}_{\mathrm{HIFtarget}}\,\mathrm{r}^{\mathrm{TTP}}_{\mathrm{HIF1A}}\left(\mathrm{r}^{\mathrm{S}}_{\mathrm{TTP}} - \mathrm{r}^{\mathrm{S}}_{\mathrm{ZFP36}}\,\mathrm{r}^{\mathrm{ZFP36}}_{\mathrm{TPP}}\right)}{\mathrm{r}^{\mathrm{TTP}}_{\mathrm{ZFP36}}\,\mathrm{r}^{\mathrm{ZFP36}}_{\mathrm{TPP}} + 1}
\end{aligned}
\tag{4.4}
$$

Since R^S_{ZFP36} maintains a constant sign the correlation is decided in the coefficient for HIF-targets. Thus the type of correlation is determined by the ratio of TTP phosphorylation ($\mathrm{r}^{\mathrm{S}}_{\mathrm{TTP}}$) to TTP protein synthesis ($\mathrm{r}^{\mathrm{S}}_{\mathrm{ZFP36}}\,\mathrm{r}^{\mathrm{ZFP36}}_{\mathrm{TPP}}$). This means that for this model scenario a positive correlation is possible if the concentration of unphosphorylated protein is decreased after perturbation. Seeing that parameter realisations result in equal numbers of correlated and anticorrelated pairs this explanation might not suffice to explain the strong correlation seen in the microarray screen (cf. Fig. 4.1).

(C) Competition models Recent reports have challenged the sequestration hypothesis by demonstrating that pTTP can still bind to 3'UTR (Sandler and Stoecklin, 2008; Clement et al., 2011). Phosphorylation was repetitively observed with simultaneous upregulation of TTP target mRNA (Stoecklin et al., 2004; Sun et al., 2007). It was therefore reasoned that pTTP could adopt a more active role in mRNA regulation than assumed in the previous model. The 3'UTR of HIF-1α mRNA encompasses 1320 nucleotides containing eight class I AREs and 2 overlapping class II AREs (Rossignol et al., 2002). This leaves enough space for two hypotheses (Fig. 4.5A). (1) The first model hypothesis assumes that pTTP loses

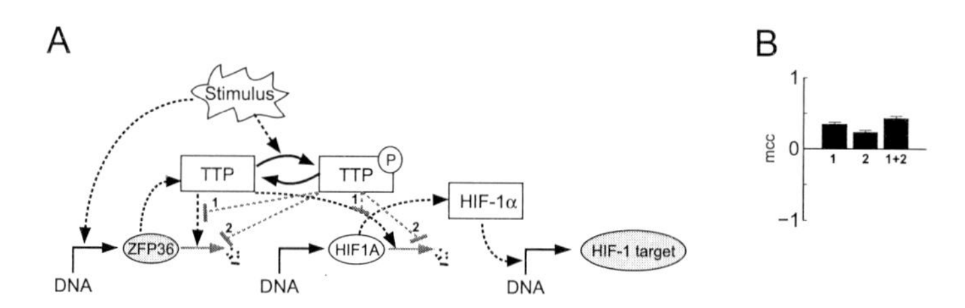

Figure 4.5 Competition models. (A) Extension of the model in Fig. 4.4 by active regulation of pTTP for two scenarios. Either (1) inactive pTTP competes with TTP for limited binding sites, or (2) binding of pTTP is not limited but stabilises the bound mRNA. **(B)** Model simulation predicts positive correlation of the two mRNAs for both mechanisms. Combining the two mechanisms further increases the positive correlation (1+2).

its regulatory activity but retains its binding potential to the 3'UTR effectively blocking TTP action by competitive binding. (2) The second model hypothesis assumes that due to the many binding sites and long UTR, both forms can bind in a non-competitive manner, but their effect on the mRNA competes by attributing a stabilising function to pTTP. Thus both mechanisms propose a role for pTTP that competes either by binding or by function with unphosphorylated TTP.

Both mechanisms produce positive correlations, with the former showing a stronger effect than the latter (Fig. 4.5B). As these two hypothesis are not mutually exclusive, the possibility of both mechanisms acting simultaneously was simulated (1+2). The combined effect produces an even stronger positive correlation (Fig. 4.5B). Therefore the competition models are more potent to qualitatively explain the positive regulation than the sequestration model, however, the exact mechanism cannot be identified without further experiments.

4.2.3 Experimental testing of model hypotheses

From the model simulations two theories, the sequestration hypothesis as well as the competition hypotheses, could serve as explanations for a positive correlation between ZFP36 and HIF-1 target expression. The main difference between these models lies in the functional role of pTTP. A decisive experimental approach to distinguish between the sequestration and the competition models would be to increase the amount of pTTP in the system and measure the response of HIF-1α mRNA. Conveniently mutant forms for a non-phosphorylatable TTP (mTTP-AA) as well as a phosphorylation-mimicking form (mTTP-PP) have been previously developed for murine TTP (Stoecklin et al., 2004). Thus it is experimentally feasible to overexpress both, the not phosphorylatable as well as the permanently phosphorylated form and monitor the effect on a target mRNA. This created the opportunity to compare the modelling results and to experimentally determine their validity.

As MRA describes steady state changes upon perturbations, effects of overexpression of TTP and pTTP on HIF1A can be readily modelled. Therefore ectopical expression of the two mutant forms was simulated by positive perturbations (+100%) of the corresponding nodes in the network and the global response of HIF-1α mRNA was recorded. Fig. 4.6A depicts box plots from 10^6 simulations of the global response coefficients from TTP and pTTP to HIF1A. When simulating the overexpression of non-phosphorylatable TTP, all models, the sequestration as well as the competition models (stabilisation and competitive binding), predict a

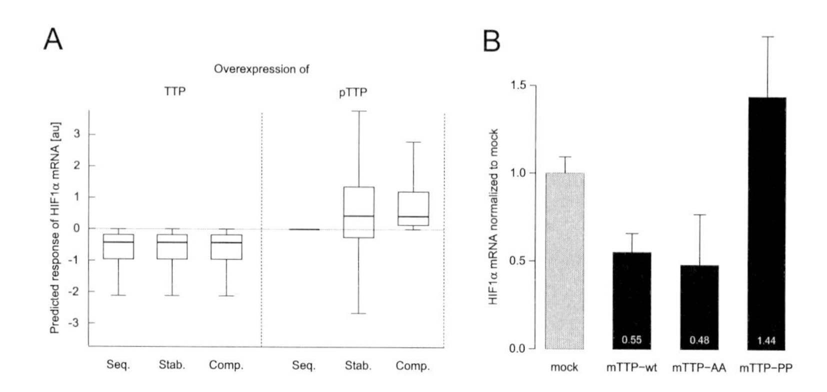

Figure 4.6 Dual role of TTP on HIF-1α mRNA regulation (A) Global response of HIF1A upon positive perturbation on either TTP or pTTP for the sequestration model (Seq.) and models with either stabilisation (Stab.) or binding competition (Comp.). All models predict a decrease of HIF1A upon TTP overexpression. For pTTP overexpression, the sequestration model predicts no change and the competition models predict largely upregulation ($n=10^6$, outliers omitted for clarity). **(B)** HIF-1α mRNA quantification in HEK293 cells when transfected with three genetically engineered variants of murine TTP: Wild-type (wt), non-phosphorylatable (AA) and constant phosphorylation-like form (PP). Whereas wt and AA downregulate HIF-1α mRNA, PP causes an upregulation thereby falsifying the sequestration model. *Figure **B** was generated by Michael Fähling.*

reduction of HIF1A in concordance with the role of TTP as a destabiliser (left panel). On the contrary, if pTTP is increased the sequestration model assumes no effect on HIF1A whereas the competition models predict an upregulation of HIF1A (right panel), stabilisation in 68% and binding competition in 100% of cases. In order to find the cause for the minor subset of negative regulations for the stabilising model, the global response coefficient was inspected.

$$R_{\mathrm{HIF1A}}^{\mathrm{pTTP}} = r_{\mathrm{HIF1A}}^{\mathrm{pTTP}} - \frac{r_{\mathrm{HIF1A}}^{\mathrm{TTP}}\, r_{\mathrm{TPP}}^{\mathrm{ZFP36}}\, r_{\mathrm{ZFP36}}^{\mathrm{pTTP}}}{r_{\mathrm{ZFP36}}^{\mathrm{TTP}}\, r_{\mathrm{TPP}}^{\mathrm{ZFP36}} + 1} \tag{4.5}$$

It can be seen that the parallel activatory role of pTTP on ZFP36 can bring about a negative feedback on HIF1A. Therefore the local response coefficient $r_{\mathrm{HIF1A}}^{\mathrm{pTTP}}$ has to be sufficiently large to counteract the increased production of TTP to achieve a positive effect on HIF1A.

These model results could now be benchmarked by experiments conducted in Fähling et al. (2012). Wild-type and mutant constructs of murine TTP were ectopically expressed in HEK293 cells and the effect on HIF-1α mRNA was compared to mock transfection (cf. Figure 4.6B). As predicted by all scenarios, overexpression of non-phosphorylatable TTP (AA) reduces mRNA levels of HIF-1α. On the other hand transfection with the phospho-mimicking TTP mutant (PP) resulted in an increase of HIF-1α mRNA, consequently refuting the sequestration hypothesis. Thus only the competition hypotheses are able to explain both experimental observations. The puzzling initial observation could thus be solved by attributing a dual regulatory role to TTP that is determined by phosphorylation status.

4.3 Discussion

This study demonstrates the applicability of modular response analysis in the field of conceptual modelling. MRA aided in generating and testing model hypothesis to determine that pTTP acts as an antagonist of TTP which could be experimentally verified. Next to an artificial system, Fähling et al. (2012) also found the stabilising effect in a more natural setting of monocytic cell differentiation by Phorbol-12-myristat-13-acetat (PMA) treatment. Western blot measurements showed that PMA produced a strong increase of phosphorylated TTP. This in turn caused stabilisation of HIF-1α mRNA by increasing the half-life from 1.3 h to 5.1 h. In this setting the phosphorylation of TTP and thus HIF-1 activation was shown to be dependent on p38 activity. TTP phosphorylation by the p38 effector kinase MK2 has been observed previously and in concordance another TTP target TNFα was found to be stabilised when p38 was activated (Tchen et al., 2004; Hitti et al., 2006; Sun et al., 2007; Sandler and Stoecklin, 2008).

To pinpoint the exact stabilising mechanism of pTTP, more elaborate experiments are required. One option would be to transfect the described mutant constructs into TTP knockout cells and monitor their effect on HIF-1α mRNA. If PP-mTTP could still increase mRNA levels, the stabilising theory would hold if not, the binding competition might serve as explanation. Unfortunately those experiments could not be done in the course of this study. Nevertheless other studies support the active stabilisation hypothesis which is thought to be mediated by the ability of pTTP to form complexes with 14-3-3 proteins that can then protect the mRNA from decay (reviewed in Brooks and Blackshear (2013)).

Since the results indicate that TTP is an important player in normoxic HIF-1 regulation, TTP status should be reviewed when encountering cancers with abnormal HIF-1 activity. Interestingly, also TTP itself represents a hub where many signalling pathways converge for target regulation (Brooks and Blackshear, 2013). For instance TTP transcription is reported to be induced by ELK1 and EGR1 in an ERK-dependent manner (Florkowska et al., 2012). Another study even showed that TTP can also regulate the ERK negative regulator DUSP6 (Bermudez et al., 2011). A recent review ascribed tumour suppressor functions to TTP as decreased TTP levels would cause an upregulation of many oncogenes such as *c-MYC*, *HIF1A* or *COX-2* (Ross et al., 2012). Based on the results learned from the modelling, not only depletion of TTP but also phosphorylation can achieve a similar, if not stronger induction. It might therefore be worthwhile to probe cancers for instances of abnormally high levels of pTTP.

The fact that phosphorylation causes a dramatic stabilising effect on TTP-bound mRNAs can serve as a good example of how to achieve an almost instantaneous increase of mRNA levels. It can even be speculated that the presumed transcriptional induction of ZFP36 observed for many stimuli may in fact originate from stabilisation by phosphorylated TTP. In this manner one could explain the positive global correlation of ZFP36 and HIF-1 activity solely by the stabilising role of pTTP. This might lead to a re-interpretation of the biological function of TTP as an activator. A phosphorylation event causes a fast upregulation of TTP targets, which due to increasing its own mRNA, is switched off as soon as the phosphorylation signal fades. Thus, TTP can produce tightly controlled bursts of mRNA levels.

This study demonstrated that next to the well known posttranslational regulators, also other mechanisms can induce HIF-1 activity and are worth to study in order to gain a complete regulation framework. Besides TTP, also other RNA binding proteins such as ELAVL1 are described to bind to HIF-1α (Galbán and Gorospe, 2009). However, no other ARE-BP, tested in here, was as strongly correlated to HIF-1 activity as TTP (see Fig. 4.1A). In order to find more regulators of HIF1A, either other candidates should be reviewed or, if the tested ARE-BPs are not regulated on mRNA level, correlation with protein abundance or -activity might yield further insights.

4.3.1 Uses and limitations of MRA for theoretical modelling

The use of the MRA framework as an analytical approach has not been extensively researched. The only previous reported use of modular response analysis in theoretical modelling has been conducted by Schulthess and Blüthgen (2011) to track information flow in biological networks of varying size. To my knowledge, this work represents the first attempt to apply MRA to conceptual modelling. The standard approach to test model hypotheses on small scale problems is to model a system of ODEs. MRA does not challenge this successful methodology but provides a convenience measure for a confined subset of problems. As the global response of MRA is defined as the steady state of a system of linear ODEs, MRA can be applied when only those states are of interest. This is of advantage especially when many parameters have to be adjusted as it takes considerably less computation time (one inversion per parameter set) than the lengthy search for steady states (thousands of iterations per parameter set).

Although in this study the methodology was used in such a manner that it produces purely qualitative results, it is capable to generate more quantitative predictions. For example, if global responses of a subset of nodes to perturbations in the network have been measured, they can be readily implemented as objective constraints for the parametrisation of the local response matrix by fitting them to the corresponding entries of R^i_j in Equation 4.2. This will reduce the available parameter space and thus result in sharper distributions of global response predictions.

Instead of rate constants and steady state concentrations of substrates, in MRA only the relative steady state change towards a perturbation can be expressed. This excludes the possibility to predict the dynamic behaviour of a network. Another prerequisite is, that MRA can only be applied to those systems where a matrix inversion for the response matrix is defined, e.g. the full rank criterion is fulfilled (Schulthess and Blüthgen, 2011). Therefore MRA can not be applied to all problems. If applicable, however, it provides a convenient short cut to quickly test modelling hypotheses, especially when feedback actions are involved.

Conclusion

A theory has only the alternative of being right or wrong. A model has a third possibility: it may be right, but irrelevant.

— Manfred Eigen, *The Origin of Biological Information*

5.1 MRA extensions enabled modelling of biological networks

Reverse engineering biological networks from incomplete and noisy experimental data is still the major challenge of contemporary network modelling. Many efforts have been devised to establish methods but so far no clear winner has emerged. From a modeller's point of view the biologically most justifiable model basis should consist of a system of ordinary differential equations. However, the experimental situation often does not allow to generate the required data for the parametrisation of dynamic ODE models. In this thesis a simplification of differential equations that only investigates the steady state characteristics of a linear ODE system was investigated: Modular Response Analysis. The MRA framework maintains the ability to quantitatively model complex network structures such as feedbacks. At the same time MRA requires only a fraction of the input a corresponding ODE model would need for sufficient parametrisation. Despite the potential of MRA as a good compromise between data requirements and modelling capacities, some shortcomings have obviated modelling attempts. These shortcomings, enlisted in the introductory chapter, were sought to be addressed in this thesis to facilitate the use of MRA in systems biology.

Since MRA is based on pre- and post perturbation steady states, which resembles the data that is regularly recorded for siRNA experiments, the method was first adapted in Chapter 2 to model gene regulatory networks. The first hurdle to overcome was the problem of how to integrate noise into the modelling framework. This was solved by decoupling the linear algebra derivation of the local response matrix from the measured perturbation data. Instead, $\mathbf{r}$ was estimated such, that the resulting $\mathbf{R}$ matches best the data-derived global response matrix, when weighted by the measurement error. This so-called maximum likelihood criterion guided a hill-climbing procedure that populated the local response matrix. Thus noisy measurements could be integrated to determine the most likely structure. The method was tested on an extensive *in silico* data set that was created to closely mimic realistic biological settings. The most critical circumstances for good performance could be identified by high signal-to-noise ratio, small network size and little uncertainty of the steady state estimations. Intriguingly high perturbations produced better predictions and nonlinearity did not affect performance gravely. When comparing the performance to other related modelling approaches on transcriptional networks a stochastic MRA approach proved to be a veritable alternative. Since the two methods perform differently well for different parameter settings they might be used according to the data properties to complement each other's shortcomings. The new ML MS MRA scheme furthermore allowed to model more complex experimental settings since also incomplete data could be handled and prior knowledge could be incorporated by fixing entries of $\mathbf{r}$ to zero. In the course of the transcriptional network modelling its functionality was demonstrated for *in silico* as well as *in vitro* data containing nodes that have only been measured but not perturbed. In this setting, RNA and protein expressions were recorded but only RNA levels were perturbed by siRNA. It was shown that including the information of not directly perturbed proteins improved the predictions, which was unmatched by the next best MRA variant.

In Chapter 3, which handled signalling networks, the experimental situation was aggravated so that in addition to measured but unperturbed nodes also some perturbed nodes could not be measured. Here, a non-identifiability analysis was implemented to determine identifiable paths and reduce the parameter set to predictive parameter combinations prior to the main MRA routine. Next, in order to include inhibitor-ligand data as well as to predict combinatorial treatments, MRA had to be capacitated to model multi-perturbations. The solution was to directly transfer the effect of perturbations into the symbolic representation of the identifiable paths. Given the low information content of the signalling data, the models were initialised with a literature-derived network structure which was then fitted and iteratively pruned off insignificant links. In this manner the local optimisation bias that has affected the genetic network reverse engineering from scratch could be largely circumvented. Overall in this thesis it could be shown that MRA can be applied to analysis tasks for both genetic as well as signalling networks.

In Chapter 4 it was demonstrated that MRA can also assist in conceptual modelling. In a study to resolve the regulatory role of TTP on HIF-1 the MRA framework was used in hypothesis testing. By iteratively populating nested minimal models the dual role of TTP could be pin-pointed and model results were experimentally tested to identify the most fit model hypothesis.

In conclusion, it can be stated that Modular Response Analysis has been shown to be a versatile base for analytical as well as predictive modelling.

5.2 Outlook I: Technical improvement plans

The here presented MRA extensions represent the first few steps in paving the way for practical modelling of biological networks. Further improvements, facilitations and alternatives can be envisioned and are delineated below.

Robust and time efficient parametrisation One of the most crucial steps in the model construction of both, signalling and genetic models, is to ensure that parametrisation is close to or at the global optimum. Since the Levenberg-Marquardt algorithm is a local optimisation routine it is advisable to repeat the model fitting with different starting values. In the signalling network approach the parametrisation was achieved by Monte-Carlo sampling, which requires dense sampling to reliably converge to the global optimum. A veritable alternative approach termed simulated annealing (Goffe et al., 1994) was tested, however, an implementation for the signalling network model was found to neither improve computation time nor overall fit. Instead, it might be worthwhile to replace Monte-Carlo sampling with a systematic sampling strategy, such as latin hypercube sampling (McKay et al., 1979; Owen, 1992). In there, the parameter space is subdivided into equiprobable hypercubes. By drawing one sample per cube it is ensured that the parameter space is best represented in each sampling run (illustration in Schelker et al. (2012)). In this manner, the ratio of optimal parameter fittings per number of total sampling runs can be increased. Thus, the initial parametrisation for the signalling network approach might require less sampling runs to reliably determine the optimal parameter values. However, this extension will only be beneficial if computation time saved by fewer runs is not cancelled out by the time needed to calculate the hypercubes. The presumed speed improvement could also allow to introduce initial value sampling in the extension phase of the genetic modelling approach and thus improve network reconstruction.

Parameter reliability estimation When modelling insufficient data, one is often confronted with the simulation dilemma. This phenomena describes the difficulty to determine whether discrepancies between experimental data and model predictions are due to an incorrect model or due to suboptimal parameter estimation (Timmer et al., 2000, 2004). In the signalling network model the goodness of parametrisation was tested by repeatedly modelling noised data and recording parameter variations. Although for the majority of parameters variation was small, future applications on larger networks might require a more elaborated parameter investigation routine. As a good candidate the profile likelihood may be used to identify (next to the already accounted structural non-identifiabilities) also practical non-identifiabilities (Raue et al., 2009). The profile for each parameter is determined by iteratively shifting the parameter value away from the optimum and recording the likelihood after fitting all remaining parameters. From the resulting profiles alone and in dependency graphs (plotting one parameter variation against another) rules about model inaccuracies and parameter insufficiencies can be spotted. This allows to assign confidence intervals for parameter estimates which facilitates the evaluation of verification experiments.

Facilitating scientific exchange Although this work demonstrates the use and applicability of MRA in various biological fields, MRA has not yet found access to the wider modelling community which relies on Boolean logic, Bayesian and Ordinary Differential Equation-based models (Novère, 2015). This work demonstrates that MRA combined with the proper extension can represent an alternative methodology. To facilitate its use for systems biology it is currently planned to provide the MRA framework as a precompiled library in the widely accessed open source-based programming language R [1]. Making the methodology accessible to the wider public will certainly lead to discussions on various model assumptions such as the dual role of the inhibitor and might give impulses for further developments.

Comparison to other statistical approaches The here introduced framework is only one of many putatively reasonable extensions of MRA and newly developed approaches might even provide a higher potential for certain application types. Shortly after the publication of the signalling network model Santra et al. (2013) published a parallel MRA approach based on Bayesian inference. Their methodology called Bayesian Variable Selection Analysis (BVSA), aims to predict a binary network structure by separately calculating the input probabilities for each measured node from all other nodes. BVSA provides a solution for noisy data and data sets where not all measured nodes were perturbed. They demonstrate in an

[1] http://www.r-project.org/

impressive array of *in silico* signalling and genetic networks as well as on an *in vitro* data set that BVSA seems to offer better prediction statistics than MC MRA as well as the genetic ML MS MRA algorithm introduced in here. Additionally, although applying a Markov Chain Monte Carlo (MCMC) sampling to calculate the posterior distributions of the network, the algorithm runs time efficient and seems to pose a useful alternative for small and medium networks. However, the current approach is severely limited as only binary network structures are inferred, giving no information about strength or activation type. Furthermore, the method can not model data with perturbed but not measured nodes nor data with combinatorial perturbations. Therefore the proposed methodology does not seem to be fit for practical use in systems biology, yet. However at a more advanced state the suggested superiority of model averaging over model selection should be re-evaluated and tested against the proposed approach in this thesis.

5.3 Outlook II: Future applications of MRA in systems biology

This work has demonstrated the utility of MRA in various fields of action such as reverse engineering of gene regulatory networks, signalling networks and conceptual modelling. However, the restriction to only perform well on small and medium-sized networks seems to be an inherent feature of all MRA approaches attempted so far (Santra et al., 2013). Therefore, the current version of MRA does not represent an alternative to whole genome modelling approaches. In addition, the recent progress in genetic high-throughput technologies spearheaded by the advent of next generation sequencing allows more and more to directly infer interactions on virtually all relevant genetic regulatory layers (van Dijk et al., 2014), on both, DNA (Plongthongkum et al., 2014) and RNA level (König et al., 2011). This may shift the computational focus away from coarse grain modelling of transcription factor networks to more detailed modelling including more regulatory structures, e.g. histone modifications, RNA-binding proteins and selective translation (Bolouri, 2014).

In conceptual modelling MRA will remain useful, alas, its simplification assumptions and restriction to only model steady state behaviour might at times be too restrictive.

Signalling network research, however, has slowly progressed and even now many processes and interactions are only rudimentarily understood. Due to the many unknown interactions and the highly interlacing structures, already relatively small

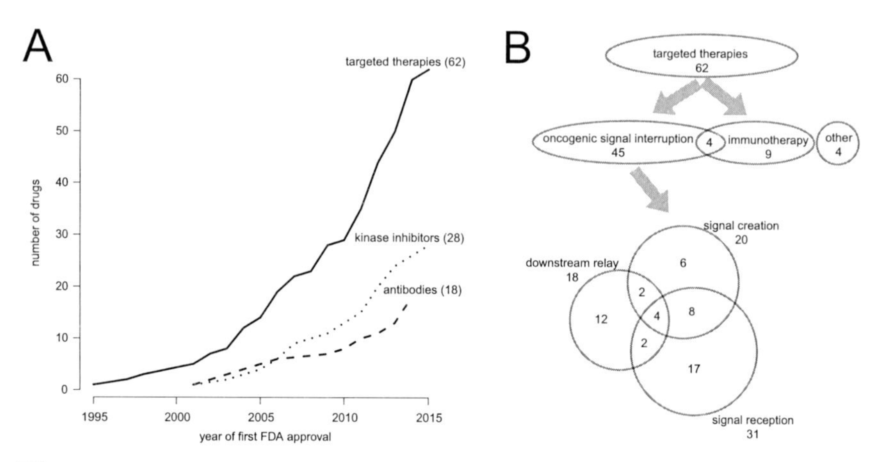

Figure 5.1 FDA-approved targeted therapy composition (**A**) Time line of number of FDA-approved drugs classified as targeted therapeutics. Development of the two major subgroups of drug types, kinase inhibitors and antibodies, are shown in pointed and dashed lines, respectively. Note that since only data for the first three months of 2015 are included the steep increase in approved drug number will likely continue. (**B**) Classification of the drugs according to their main therapeutic action. For detailed information see Suppl. Tab. D.1.

networks evade intuitive interpretation. More importantly, it is considered that the signalling state of the cell represents the best proxy for the action state of the cell and may thus be most conclusive about the phenotype. Therefore the greatest potential of MRA resides in analysing signalling networks and in there with the prospect of researching diseases such as cancer.

5.3.1 Analysing signalling in cancer research

Targeted therapy-oriented cancer research is firmly focussed on cell signalling. This is reflected by the fact that 49 out of 62 targeted therapeutics, that have been approved by the U.S. Food and Drug Administration (FDA) as of March 2015, interfere with oncogenic signalling (see Figure 5.1). It can be noticed that the progress of targeted therapies to reach the clinics has accelerated since 2010. In order to ensure the rapid pace of developing more potent treatments with less adverse effects, network analysis in general (Kholodenko et al., 2012) and MRA in particular can contribute in various aspects.

Study re-wiring effects upon induction of carcinogenic mutations To be able to devise the best treatment strategy for currently 'undruggable' oncogenes such as RAS and to recognise inherent resistance mechanisms, it is important to understand the shift in network connectivity in response to a mutation event. As outlined in the signalling chapter, an ideal approach would be to characterise the signalling network of isogenic cell lines that differ by a single somatic mutation. Among the different experimental possibilities, an inducible system would yield the best comparability between wild type and mutant cells, as clonal effects would be cancelled out. The re-wiring of the signalling network can then be described by feedback-capable modelling approaches such as MRA. Thus the oncogene introduced vulnerabilities can be pin-pointed and corresponding treatment strategies that exploit synthetic lethality (Kaelin, 2005; Brough et al., 2011) can be devised. In an analog fashion the loss-of-function effect of tumour suppressors can be exploited, potentially increasing the number of treatment options.

Uncover mechanisms of acquired resistance The euphoria upon the initial success of targeted therapies has been overshadowed by the high rate of relapses often occurring within several months. This effect has been described as acquired resistance (Holohan et al., 2013). In order to develop appropriate follow-up treatments as well as to find ways to prevent relapse, possible resistance mechanisms have to be identified. Similar to the previous paragraph a first step would be to study the difference in signalling before and after resistance occurred in a cell. The insensitisation of originally sensitive cell lines by long-term culture with sublethal drug doses has been demonstrated (e.g Mader et al. (2014)) and can be applied to derive corresponding cell line models. By quantitatively comparing the signalling of the resistant cells with the sensitive parental cell line the mechanisms of resistance may be found and preventive measures can be developed. Another common type of acquired resistance is caused by the genetic heterogeneity of the tumours, where resistant subpopulations take over the tumour (Gerlinger et al., 2012). In this case follow-up treatments can be devised by analysing the signalling of patient-derived cell models of acquired resistance (Crystal et al., 2014).

Determine potent combinatorial treatments An often cited approach to overcome emergence of acquired resistance is to replace single agent targeted therapeutics by a synergistic combination therapy (Straussman et al., 2012; Miller et al., 2013; Rad et al., 2013; Lai et al., 2014; Misale et al., 2014). In spite of these indications, only one small kinase inhibitor combination has been approved by the FDA to date, i.e. RAF+MEK inhibitor in melanoma (Menzies and Long, 2014)). With only 28 kinase inhibitors approved in total, the combinatorial research field is comparatively young (Figure 5.1). Furthermore, several combinations are currently tested in clinical trials. Among them the MEK+RAF+EGFR therapy[2] for colorectal cancer which was proposed in this work. However, the greatest potential resides in identifying drug combinations where single treatments have shown little effect. The majority of approved targeted therapeutics that interrupt oncogenic signalling interfere with oncogenic signal creation or reception (cf. Figure 5.1). In contrast, only 12 drugs solely target downstream relay which might hint to buffering by feedbacks and crosstalks which can only be overcome by an appropriate combinatorial treatment (Klinger and Blüthgen, 2014). Due to combinatorial explosion, exhaustive *in vitro* testing for synergy is impossible for the plethora of candidate drugs. Here, model predictions from single perturbation data can filter out the most promising combinations (Klinger et al., 2013; Molinelli et al., 2013).

Exploring potential of signalling network biomarkers Drug discovery is concentrating on single mutations that occur frequently in distinct tumour entities, often ignoring the combinatorial effects of additional mutations. Also, a large number of tumour-related mutations exist, that never occur in high frequency and therefore lack proper treatment options (Luo et al., 2009). Despite the diversity on the genome level, it can be speculated that the combined mutational effect on the key mediators in the executive layer might be less diverse. This calls for a search for specific characteristics in the signalling network which may respond to the same type of treatment. MRA could help to identify those signalling biomarkers by comparing and analysing network parameters of a large cell line panel with diverse mutational background. If successful, one could then try to answer the original question by mapping signalling phenotype back on the corresponding set of mutational conformations.

Decipher cell-cell communication The translation of cell culture-derived treatment options to clinical application is often hampered by the disregard of the interaction between the tumour and surrounding stromal cells (Trédan et al., 2007). Tumour promoting stroma consists of angiogenic cells, immune cells and cancer-associated fibroblasts also termed tumour microenvironment (Hanahan and Coussens, 2012). Of particular interest is the interaction of immune cells that do not attack and even support tumour growth (Whiteside, 2008). As the extended MRA is now able to model multiple perturbations, it might be used to reconstitute the signalling state that is produced by the interaction. First, the models should

[2] `https://clinicaltrials.gov/show/NCT01750918`

be trained on a large scale data set of defined perturbations for both, tumour and immune cells. Then, the signalling is recorded in co-culture and sought to be reproduced from the training set. Thus the role and relevant signalling paths might be elucidated in a mechanistic manner which can provide means to interfere with intercellular signalling or to reactivate immune cells. In this respect MRA could also assist in the development of immune therapy approaches, the second largest class of targeted therapies (cf. Fig. 5.1).

5.4 Closing remark

This work has attempted to create use-cases for the long-neglected modelling technique MRA. Any model is always an abstraction of the truth and therefore wrong, but maybe useful (Box, 1979), which I hope I could illustrate in here. Karl Popper argued that the progress of science starts from the most fit (not necessarily most true) theory until it is refuted which brings about a new theory and so forth (Popper, 1957). This process of trial and error can be adapted to the use of mathematical models as well. Given the current state of technology and focus of research, the most fit model should be devised. Thus I propose that MRA can inhabit the feedback-incorporating niche where other models are inadequate either due to theoretical or practical reasons. This niche will persist until technology or theory further develops to allow the use of better model systems. Thus to return to the original quote of Manfred Eigen, MRA seems to produce relevant models for now, whether they are right or wrong remains to be tested, though.

The role and use of mathematical modelling is often disputed with regard to biological research. In this work I tried to illustrate that one can perceive modelling as a powerful filter that can be used for two main purposes: (i) to filter data and (ii) to filter ideas. Thus models help to better conceptualise or improve the visualisation of data in the noisy world of biology, in a similar fashion as telescopes and microscopes aid the naked eye - they help to see clearer and to focus. The interpretation of model results, however, remains a challenge of the human mind.

Appendix

Appendix for chapter: Reverse engineering of genetic networks

A 1 Parametrisation of simulated genetic networks

To simulate realistic data, the parameter choice was attempted to be guided by available literature findings. The most comprehensive dataset that could be found stems from a high-throughput study on mouse fibroblasts and encompasses the four crucial parameters transcription rate $V_{\max,R_i}$, translation rate $V_{\max,P_i}$, and RNA and protein degradation rates (d_{P_i}, d_{R_i}) (Schwanhäusser et al., 2011). In Fig. A.1A it is shown that the distribution of all four parameters can be well approximated by a log-normal distribution with the corresponding mean and standard deviation. However, the absolute numbers represent only a subset of the genome (≈ 5000 genes) and filtering out only transcription factors (as defined in AnimalTFDB (Zhang et al., 2012)) results in a list of merely 124 entries which is only a small fraction of the 1447 murine transcription factors stored in AnimalTFDB. Furthermore, there is evidence that the average rates of transcription factors are different from proteins of other function. In another murine whole genome study of Sharova et al. (2009) it was shown that mRNA half-life of transcription factors was shorter than the half-life of other transcripts. This observation has also been made in human cells (Yang et al., 2003). Therefore, the actual parameters for transcription factors might not be well represented by drawing them from Fig. A.1A. However, the underlying distribution might hold for all those parameters which can be recapitulated for mRNA decay rates of transcription factors derived from Sharova et al. (2009) shown in Fig. A.1B. In conclusion, to generate the artificial gene regulatory networks, all parameters were drawn from log-normal distribution and plugged into Eq. (2.5) on page 22. Since the actual figures are not known, absolute magnitude and variation of all production and degradation rates were set to the same value ($\mu = 1.6, \sigma = 2.1$) and the constants K were set to half of that value ($\mu = 0.8, \sigma = 1.1$)[1].

[1]Using rates from Schwanhäusser et al. (2011) resulted in a similar distribution of r_j^i's.

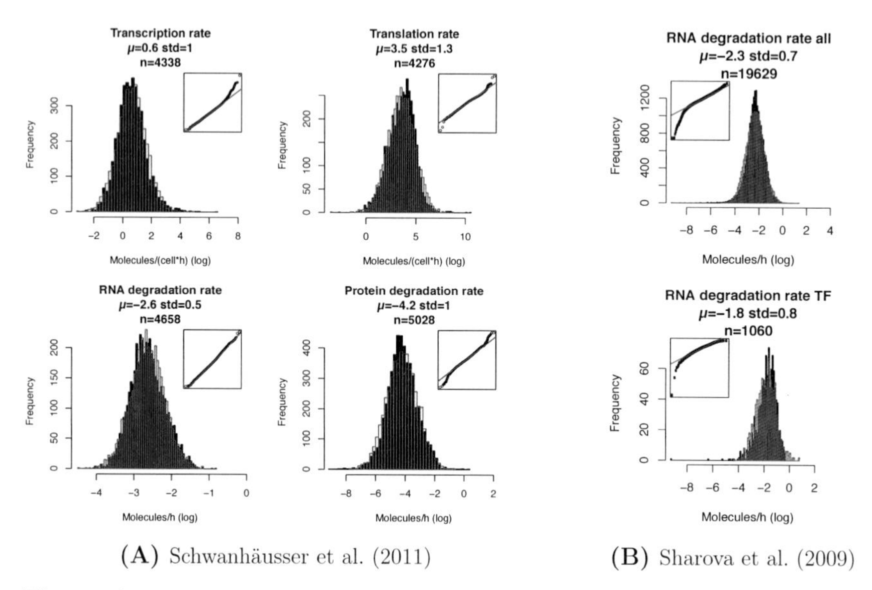

(A) Schwanhäusser et al. (2011) (B) Sharova et al. (2009)

Figure A.1 Production and degradation rate of proteins and RNA can be approximated by log-normal distributions Histograms of experimentally derived rates (black) from (**A**) mouse fibroblasts and (**B**) mouse embryonic stem cells compared to randomly drawn samples from log-normal distributions with the corresponding mean and standard deviation (red). INSETS: Quantile plots in (A) show slight deviation from log-normal distribution close to the limits which might be due to technical limitations. Quantile plots in (B) exhibit deviation from log-normal distribution close to lower limits indicating an artefact for long half-lives owing to lack of sufficiently long time points in the experiment. Data source is indicated in the respective figure caption. Transcription factor definition taken from AnimalTFDB (see main text).

Reassuringly, in Fig. A.2A it was found that the steady state correlation of the unperturbed RNA and protein species for the standard transcriptional network set correlates by 42%, resembling experimental findings. The default external and structural network settings were set to network size of 10, link density of 20%, perturbation strength of –50% of original V_{max}, and noise to 20% of median signal value.

In contrast to experimental data, in simulated networks the local response coefficient can be approximated. To achieve this, the unperturbed steady state was systematically perturbed by a minimal amount (1%) and allowed to propagate for one iteration step. From this, the local response coefficient can be approximated by normalising the difference of the pre- and post-perturbed state by the initial perturbation strength and afterwards by scaling each row vector r_i by the directly

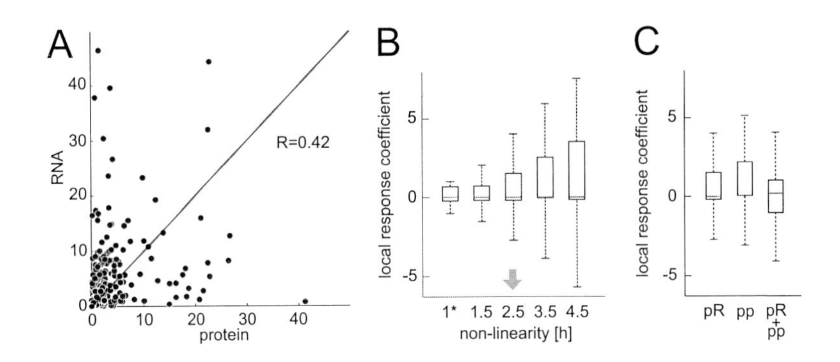

Figure A.2 Test set characteristics (**A**) Correlation of RNA to protein steady state of 100 ten-node transcription factor networks with default parameters. (**B**) Boxplpots of approximated local response coefficient strengths for default transcription factor networks with varying hill coefficient h. The x-axis displays the average h with a fixed standard deviation (0.7). Asterisk denotes the linear case were the hill-like coefficient was fixed to 1. The arrow points to the default setting. (**C**) Approximated local response coefficient distributions in default network settings for different interaction scenarios: transcriptional network (pR), postranslational networks (pp), and a mix of both (pR+pp).

perturbed entry $-r_i^i$. The exact values for the model parameters were sought to be adjusted such that simulated r's were on average considerably far from zero. Among the tested parameters only the hill coefficient was found to produce larger local response coefficients with higher mean values (cf Fig. A.2B). It can be seen that especially for positive coefficients non-vanishing values are obtained and results larger than one are reached for both signs. However, as cooperativity also implies a higher degree of nonlinearity a value that produced sufficiently large local response coefficients and only medium cooperativity was chosen for h ($\mu = 2.5, \sigma = 0.7$).

In Fig. A.2C the distribution of r's for the default network is shown for all three network scenarios. Whereas for the mixed model (pR+pp) a symmetric distribution can be achieved for both transcriptional (pR) as well as postranslational networks (pp) this distribution is skewed to positive values. However, a fraction of 25% negative local response coefficients was termed acceptable as in realistic biological networks more positive than negative interactions are expected.

A 2 Evaluation statistics

The methods were scored based on count statistics. Four categories were recorded: (i) number of links correctly recovered with matching sign (TP), (ii) number of missed links (FN), (iii) number of falsely predicted links (FP), and (iv) number of links correctly predicted to be absent (TN). In the count data trivial connections, i.e. the diagonal of the matrix, or translation, were not considered. Furthermore, forbidden interactions such as mRNA-mRNA interactions were excluded as well. Depending on which network scenario was in place, the not modelled interaction was excluded as well (i.e. pp regulation for pR scenario and *vice versa*). As a gold standard for link presence and sign the local r approximation described in Suppl. Section A 1 on page 113 was considered. In this manner it was ensured that those links are filtered out that have been specified in the structure but cannot be recapitulated from the data. The resulting count data were then used to calculate **sensitivity** ($\frac{TP}{TP+FN}$), **precision** ($\frac{TP}{TP+FP}$), and **Matthews correlation coefficient (mcc)** that were used for evaluation. Mathews correlation coefficient was selected due to its described superiority among single number classifiers (Baldi et al., 2000) and is defined as:

$$mcc = \frac{TP \cdot TN - FP \cdot FN}{\sqrt{(TN + FP)\,(TP + FN)\,(TN + FN)\,(TP + FP)}}. \tag{A.1}$$

Belonging to the same type, mcc shares many characteristics of correlation coefficients. Ranging between 1 and -1 the extremes denote perfect prediction and inverse interpretation of link nature. A coefficient around 0 usually denotes random predictions which are due to lack of information such as trying to predict postranslational networks from RNA perturbation data or due too an insufficiently low signal-to-noise ratio. In the current network study random predictions often exhibit a slightly negative mcc. Due to the maximum likelihood approach model predictions for uninformative models will produce only few significant links. Due to the sparse network definition random guesses will make less true positive than false positive predictions. This imbalance can normally be countered in (A.1) by weighing the positive predictions with the corresponding negative predictions. In case of TP=0 and FP>0 this balance can only work on FP and will nevertheless produce slightly negative correlation coefficients. Therefore small negative values of mcc from predictions with low true link numbers will be regarded as random.

A 3 Complete parameter benchmark results of ML MS MRA

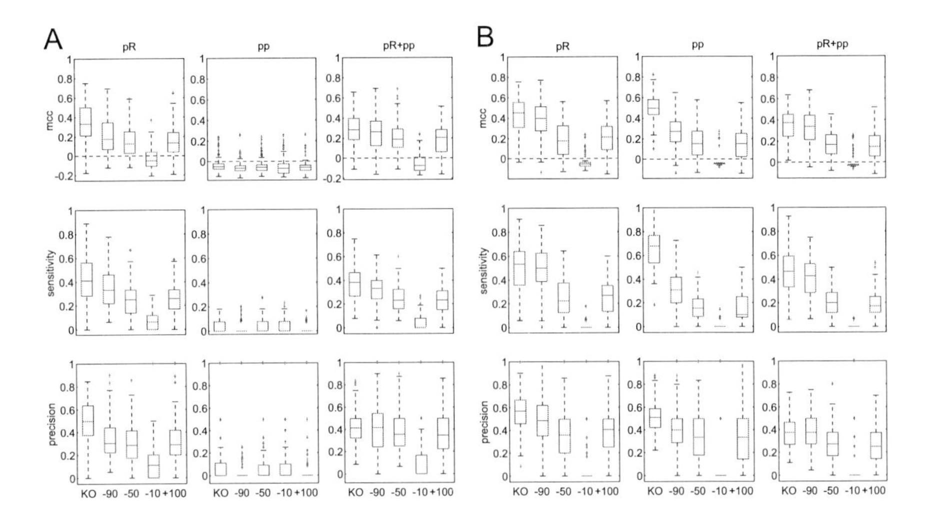

Figure A.3 Effect of perturbation variation on prediction Statistical evaluation of the prediction of ML MS MRA on 100 simulated ten node networks for different levels of perturbations based on (**A**) RNA data and (**B**) RNA+ protein data. Initial perturbation changes are indicated as percent change of the original V_{max} and include a knockout (KO) and overexpression scheme as well as three knockdown scenarios. The dashed line indicates 0 for Matthews correlation coefficient (mcc).

In the following the complete test set benchmarks for the five varied external parameters, perturbation strength, noise, non-linearity, network size and sparsity are visualised. For each scenario all predictions for the three network types containing only transcription factor interactions (pR), posttranslational interactions (pp) and both (pR+pp) are shown. Since for many biological settings such as microarray or RNA-Seq studies often only mRNA data is available, the performance when only providing mRNA data is also investigated. In general it can be seen that the method performs worse on RNA only data, especially for posttranslational modifications. This can be explained with the fact that RNA + protein contains more information and by acknowledging that posttranslational modifications can only be indirectly inferred from RNA data.

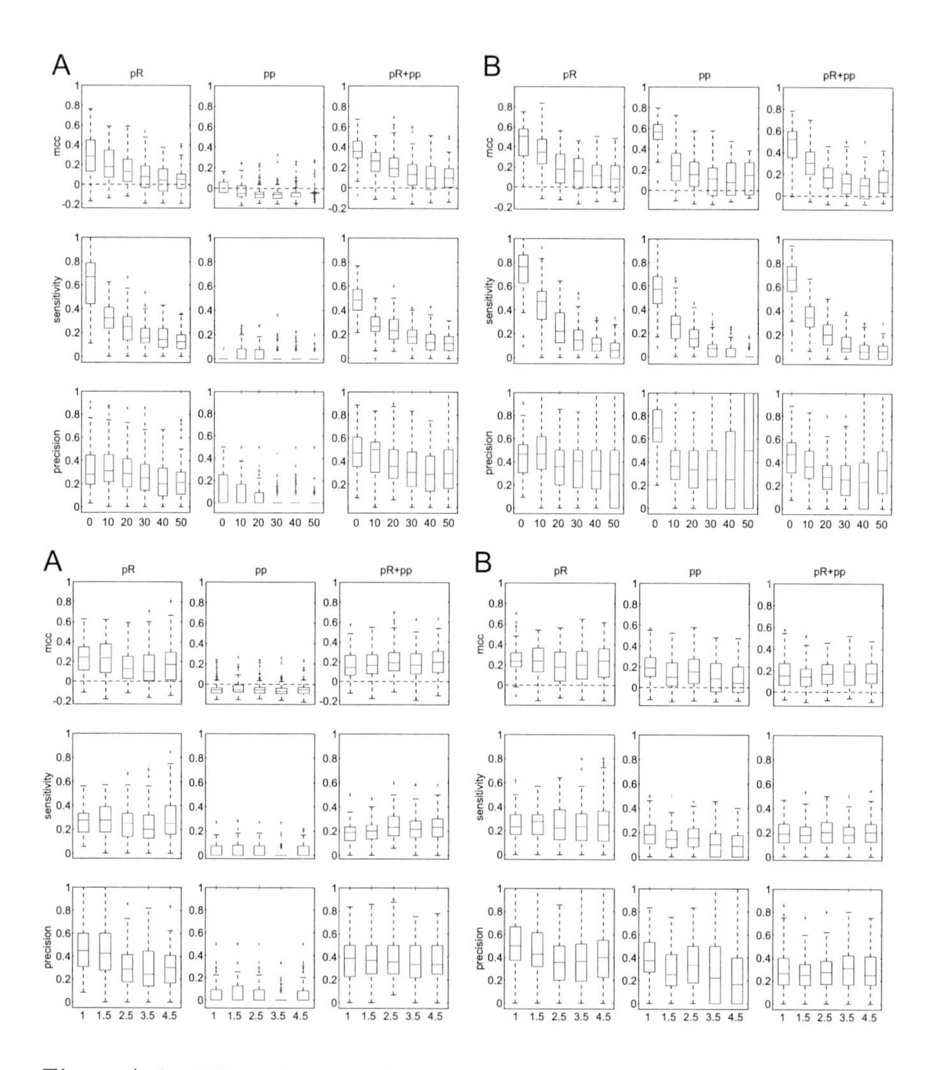

Figure A.4 Effect of noise and nonlinearity Statistical evaluation of the prediction of ML MS MRA on 100 simulated ten-node networks for (TOP) different levels of noise (percent) or (BOTTOM) various degrees of nonlinearity performed on (**A**) RNA data and (**B**) RNA+ protein data. To simulate different grades of nonlinearity the parameter h in Eq. 2.5 on page 22 was modulated. The x-axis depicts the mean hill coefficient (h) with constant variation of 0.7. To show the difference to a 'linear' model, a further test set with constant $h = 1$ was simulated (first columns).

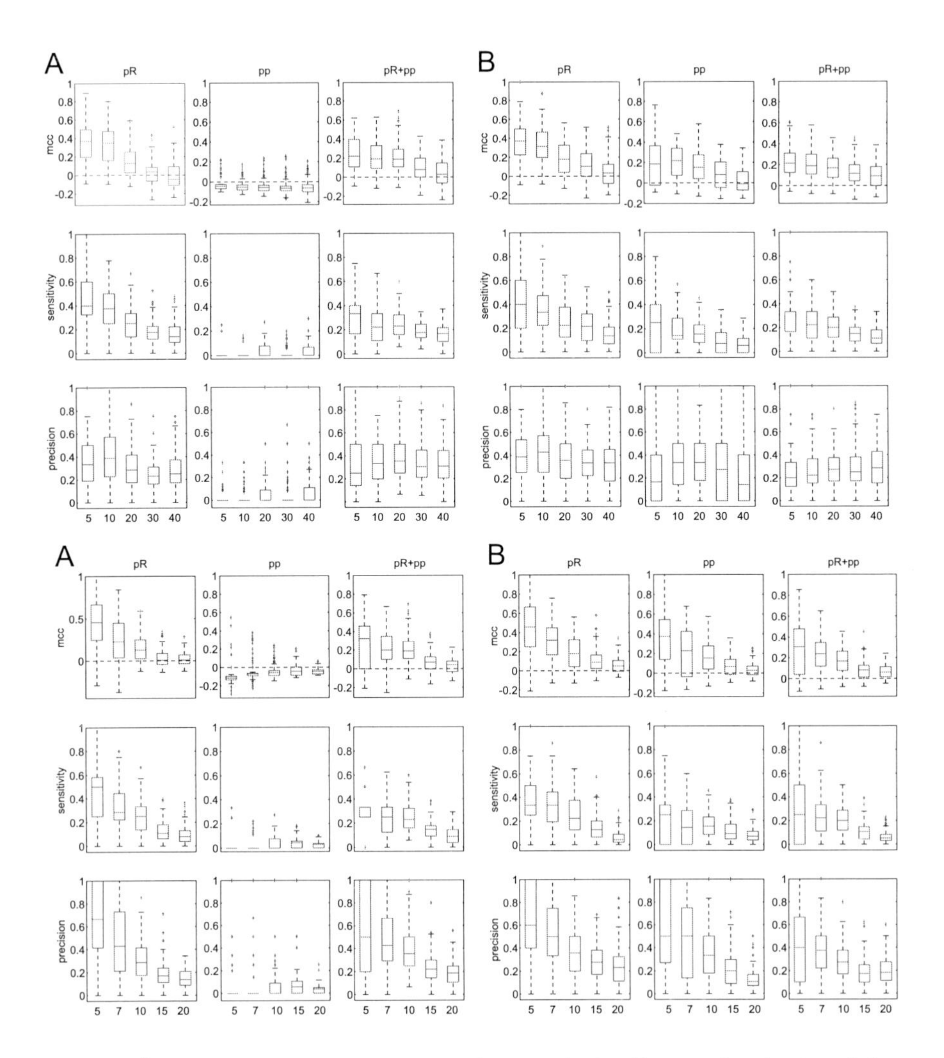

Figure A.5 Effect of link density and network size Statistical evaluation of the prediction of ML MS MRA on 100 simulated networks with (TOP) different link densities (percent) or (BOTTOM) network sizes performed on (**A**) RNA data and (**B**) RNA+ protein data.

A 4 Alternative steady state prediction methods

The alternative approaches that were tested against the proposed Maximum Likelihood Model Selection MRA (ML MS MRA) are briefly described. For comparability the same number of links, that have been recovered by ML MS MRA were taken which were selected according to the method's inherent scoring scheme.

Plain MRA The original MRA was applied as proposed by (Kholodenko et al., 2002) which is outlined in Fig. 1.2 on page 11 in the main manuscript. First, the global response matrix was calculated from the steady states before and after perturbation. From the global response matrix the local response matrix was calculated by using Eq. 1.5 on page 10. The confidence of links was assumed to be related to the absolute strength and consequently the largest local response coefficients were kept for the statistical comparison. As the plain MRA requires all nodes to be measured and perturbed, network prediction was restricted to RNA data only.

MC MRA Monte Carlo MRA (MC MRA) was essentially implemented as described in Santos et al. (2007). The approach in there is based on the theoretical reasoning of Andrec et al. (2005) which is summarised below. The central MRA equation postulates that the matrix product of the local and global response matrix result in a matrix with the negative perturbations p_i on the diagonal and zeros for all off-diagonal entries. The entries of the off-diagonal are thus defined as:

$$\sum_k r^i_k R^k_j = 0 \quad (i \neq j) \quad . \tag{A.2}$$

This means that all row vectors r^i are perpendicular to all column vectors R_j as long as $i \neq j$. This relationship can be reformulated into a linear equation system:

$$r^i_i R^i_j + \sum_{k \neq i} r^i_k R^k_j = 0 \quad \overset{r^i_i=-1}{\Longleftrightarrow} \quad \sum_k r^i_k R^k_j (1 - \delta_{ik}) = R^i_j \qquad (i \neq j)\,. \tag{A.3}$$

With δ_{ik} denoting the Kronecker delta which is 1 for $i = k$ and 0 otherwise. This equation system is solved by the total least squares solution (TLS) separately for each row vector r^i. This can be achieved by applying singular value decomposition to each column vector R_j and scaling by the terminal right singular vector so that $r^i_i = -1$.

The link probability was estimated from repeated simulations ($n = 10^6$) by adding normally distributed noise with $\mu = 0$ and standard deviation taken from the error estimation (same as for ML MS MRA see Suppl. Section A 6 on page 125). Each of these simulations resulted in a different realisation of the local response matrix, $\tilde{\mathbf{r}}_\mathbf{i}$. The final local response coefficients in $\mathbf{r}$ are calculated from the median of the corresponding entries of the $\tilde{\mathbf{r}}_\mathbf{i}$'s.

Ranking of links was defined by the confidence measure. The confidence of each link was assessed by inspecting the percentage of simulations in which the inferred link had the same sign as in the respective r^i_j. Links with absolute value of r^i_j smaller than 10^{-5} were treated as absent.

To set forbidden interactions to zero (e.g. mRNA-mRNA interactions), the relevant column(s) in the global response matrix were removed before applying singular value decomposition and the positions filled with zeros when the r^i's were reassembled. Therefore this method allowed the prediction from RNA data alone as well as RNA plus protein data.

NIR Network Identification by multiple Regression was applied according to its first description by Gardner et al. (2003). The initial assumption of NIR is essentially the same as in MRA. A perturbation vector $\mathbf{u}_i$ can be linearly approximated by the steady state responses in nodes $\mathbf{x}_j$ when weighted by a matrix $\mathbf{A}$ which represents the network connectivity:

$$-\mathbf{u}_j = \mathbf{A}\mathbf{x}_j \quad . \tag{A.4}$$

If all nodes are systematically perturbed this system can in principle be solved but it is argued by Gardner et al. (2003) that noise causes the system to be underdetermined. Assuming that M experiments have been conducted, an overdetermined situation is then sought to be re-established by limiting the maximal number of input edges per node k to be below M. To determine the best vector of input combinations $\tilde{\mathbf{a}}^i$ of maximal k nodes on node i the minimal sum squared error (SSE) is estimated by ordinary least squares:

$$\tilde{\mathbf{a}}^i = \underset{\tilde{\mathbf{a}}^i \in \mathbb{R}}{\arg\min}\, SSE^k_i(\tilde{\mathbf{a}}^i) = \left(\tilde{\mathbf{X}}^{\mathrm{T}}\tilde{\mathbf{X}}\right)^{-1}\tilde{\mathbf{X}}^{\mathrm{T}}u^i \tag{A.5}$$
$$\text{with} \quad SSE^k_i(\tilde{\mathbf{a}}^i) = \sum_{l=1}^{M}\left(u^i_l - \tilde{\mathbf{a}}^{i\mathrm{T}}\tilde{\mathbf{x}}_j\right)^2 \quad .$$

In here $\tilde{\mathbf{X}}$ is a $K \times M$ submatrix of steady state ratio matrix $\mathbf{X}$ and vector $\tilde{\mathbf{a}}^i$ contains the k non zero entries of row i in matrix $\mathbf{A}$.

The matlab implementation of the NIR algorithm, was downloaded from `http://dibernardo.tigem.it` and provided with the corresponding entries. As measured input the global perturbation of node j is quantified as the deviation from parity of the pre- $x_{(0)_i}$ and post-perturbation steady state $x_{(1)_{ij}}$:

$$X_j^i = \frac{x_{(1)_{ij}}}{x_{(0)_i}} - 1 \quad . \tag{A.6}$$

In here perturbation strengths were given as the relative change of V_{max} before and after perturbation and no uncertainty of perturbation was assumed. As a first indicator, a method-inherent F-test was considered which estimates for each node whether any combination of k input links significantly improves the sum squared errors when compared to no input on that node ($\alpha = 0.05$). If no combination passes the threshold all input links of that node were set to 0. Otherwise the most significant parametrisation was kept. In case that for different k's several significant improvements could be achieved, the solution that resulted in the best Matthews correlation coefficient was taken. For this a subset of links in **A** was taken, only including as many links as were included in the test set.

The ranking of links was achieved by ordering after signal-to-noise ratio. The signal-to-noise-ratio was determined by dividing **A** by the variance of the weights **sA** estimated from the covariance matrix of **A** (method inherent). In order to derive those weights measurement errors had to be provided wich where derived by Gaussian error propagation of Eq. A.6:

$$\begin{aligned}
\epsilon_{X_j^i} &= \sqrt{\left(\frac{df(X_j^i)}{dx_{(1)_{ij}}}\epsilon_{x_{(1)}{}_j^i}\right)^2 + \left(\frac{df(X_j^i)}{dx_{(0)_i}}\epsilon_{x_{(0)}{}_i}\right)^2} \\
&= \sqrt{\left(\left(\frac{x_{(1)_{ij}}}{x_{(0)_i}} - 1\right)'_{x_{(1)_{ij}}}\epsilon_{x_{(1)}{}_j^i}\right)^2 + \left(\left(\frac{x_{(1)_{ij}}}{x_{(0)_i}} - 1\right)'_{x_{(0)_i}}\epsilon_{x_{(0)}{}_i}\right)^2} \\
&= \sqrt{\left(\frac{1}{x_{(0)_i}}\epsilon_{x_{(1)}{}_j^i}\right)^2 + \left(\frac{-x_{(1)_{ij}}}{x_{(0)_i}{}^2}\epsilon_{x_{(0)}{}_i}\right)^2} = \underline{\underline{\sqrt{\frac{\left(x_{(0)_i}\epsilon_{x_{(1)}{}_j^i}\right)^2 + \left(x_{(1)_{ij}}\epsilon_{x_{(0)}}\right)^2}{x_{(0)_i}{}^4}}}} \quad .
\end{aligned}$$

Sorting for $|\frac{A_i^j}{sA_i^j}|$ provided the most confident links that were kept for comparison with the MRA methods as well. The available matlab routine only handles fully perturbed systems therefore RNA only data were evaluated.[2]

[2]In Gardner et al. (2003) a way to model unperturbed nodes is stated but seems to be unrealised.

A 5 Extended modelling results for DREAM 4 challenge 2

The specific DREAM challenge was to predict network structures from five different perturbation data. In the main text (Section 2.4.4 on page 35) it is delineated that all MRA variants can produce third-ranked results when predicting only from knockout data and six replicate measurements of the pre-perturbed steady state $x(0)$. In Fig. A.6 two statistical measures, which weight the ranking of each link position is shown for all three variants. The ROC curve depicts the true positive rate ($TPR = TP/TP + FN$) in dependence of the false positive rate ($FPR = FP/FP + TN$). Better than random predictions are indicated by ROC curves that are above the identity line. The P-R curve denotes the dependency of the precision on the recall (i.e. TPR). In this case a curve that lies higher than the negative identity would indicate a better prediction (i.e. upper right quadrant). The curves can be summarised by calculating the area under the curve called AUROC and AUPR whose value ranges between 0 (wrong) and 1 (perfect prediction).
The resulting scores are comparatively high so that for three of five networks at least one of the methods achieves an AUROC score higher than 0.9. Even though the AUPR is in general worse, for three networks a score of at least 0.7 ist achieved.

The depicted curves are a good complement to study the statistical behaviour of the methodologies that might be missed when using only single number classifiers. In here transition points can in principle be visualised and rules for the best cutoff strategy might be learned. However, from these five networks no clear rules for a good balance between low false positives and a good network representation are evident. Instead, the network predictions have to be evaluated also for their practical use by viewing the statistics for the high confidence links only. In Fig. A.6 the circles denote the final point on the curve when only significant links are taken. For ML MS MRA and MC MRA significance levels were set to 0.05 of their inherent statistics and for plain MRA r's larger than 0.2 were considered (following Stelniec-Klotz et al. (2012) definition of irrelevant links). Even though the scores for the area under the curve (in brackets) are hardly changed, precision is found to be lower than 60% in three networks for each of the two MRA extensions and in all five networks for plain MRA.

In order to asses this in more detail in Tab. A.1 the single value statistics that have been used for the *in silico* benchmarking are applied (left columns). It can be that in contrast to the overall score the mcc indicates a scheme that splits honours between MC MRA and ML MS MRA. In two cases MC MRA produces

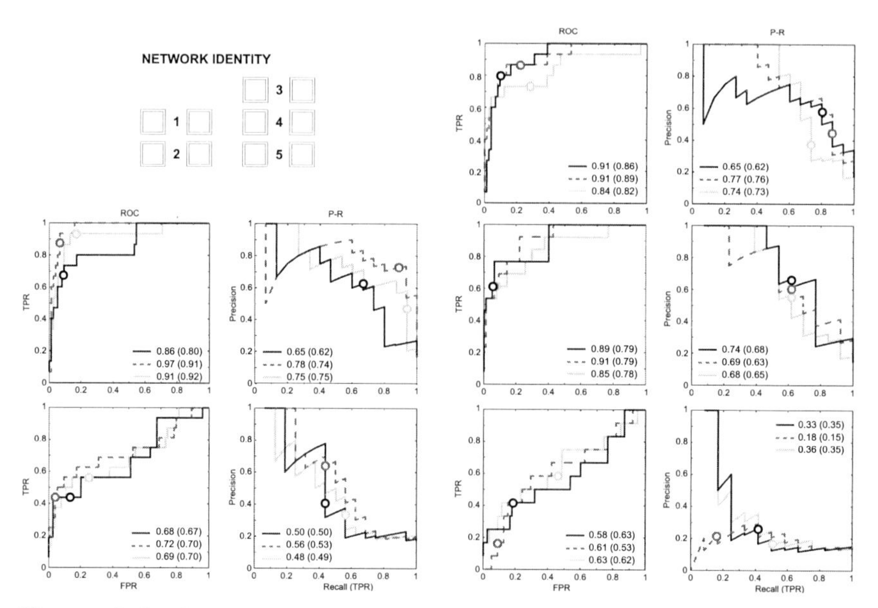

Figure A.6 Complete DREAM 4 challenge 2 outcomes for MRA variants (**A**) Receiver operation curves (ROC) and precision recall curves (P-R) for ML MS MRA (black), MC MRA (dark grey dashed) and plain MRA (grey) on five different *in silico* networks. The respective area under the curve scores (AUROC and AUPR) are indicated in the figures. The number in brackets denote the score when only significant links were considered ($|r| >= 0.2$ for plain MRA and $\alpha \leq 0.5$ otherwise), which corresponds to curves ending at the depicted circles.

much better results (networks 1 and 2) and for two networks ML MS MRA would be the better choice (networks 3 and 5) whereas for one network performance is comparable (network 4). Plain MRA with the used cutoff can not serve as a good alternative as it seems to be often not stringent enough. Three problems of ranking MRA by strength can be envisioned. First, the strength distribution varies from network to network which is evident in the five networks and leads to a span of 14 links in network 4 and 42 links in network 5. Second, the influence of r's of equal strength on the network is different for r's connecting to end nodes and r's connecting nodes inside the network. And third, due to heterogeneous influences of the error in measurements the probabilities for link existence might be different for links of equal strengths. When providing MC MRA and plain MRA with the ML MS MRA link number cutoff (Tab. A.1 right) mcc is always better for plain MRA and twice better for MC MRA predictions. This indicates that the maximum likelihood cutoff is a good predictor for the number of links to

Table A.1 Evaluation of practically feasible network predictions from DREAM 4 Challenge 2 Shown are the statistics sensitivity, precision, and Matthews correlation coefficient (mcc) for five networks reverse engineered by three different variants of MRA. The significance threshold was calculated in two ways, once for the inherent cutoff ($\alpha \leq 0.05$ for MC and MLMS and $|r| > 0.2$ for plain MRA) and once for an equal number of links as produced by ML MS MRA.

network	method	Inherent cutoff			MLMS cutoff	
		Plain	MC	MLMS	Plain	MC
1	sensitivity	0.93	0.87	0.67	0.67	0.80
	precision	0.50	0.76	0.63	0.67	0.80
	mcc	0.60	0.77	0.57	0.60	0.76
2	sensitivity	0.56	0.44	0.44	0.50	0.56
	precision	0.33	0.64	0.41	0.47	0.53
	mcc	0.27	0.45	0.30	0.37	0.44
3	sensitivity	0.73	0.87	0.80	0.73	0.80
	precision	0.34	0.43	0.55	0.50	0.55
	mcc	0.35	0.51	0.58	0.51	0.58
4	sensitivity	0.62	0.62	0.62	0.54	0.54
	precision	0.57	0.62	0.67	0.58	0.58
	mcc	0.52	0.55	0.58	0.49	0.49
5	sensitivity	0.58	0.17	0.42	0.42	0.42
	precision	0.16	0.22	0.26	0.26	0.26
	mcc	0.07	0.09	0.20	0.20	0.20

consider. From the six replicates of the unperturbed steady state (wildtype + five zero time points from the time series data) the noise level of the networks could be approximated. The Coefficients of Variation (CV) were $\approx 12\%$ for networks 3 and 4, 14.5% for network 1 and 19% for the second and fifth network. Although the higher noised networks produced generally worse performances the assertion that low noise prefers MC MRA and high noise ML MS MRA does not hold in here. This might be due to the low sampling size or the inaccurate error estimation of other not repeated data points.

A 6 Assessment of two estimates of the global response matrix

In Kholodenko et al. (2002) two ways to approximate the global response matrix **R** from experimental data are given. Next to the original definition $R^i_j = \ln\left(\frac{x_{(1)_{ij}}}{x_{(0)_i}}\right)$ the approximation $\mathbf{R}^i_j = 2\left(\frac{x_{(1)_{ij}} - x_{(0)_i}}{x_{(1)_{ij}} + x_{(0)_i}}\right)$ is suggested as an experimentally more feasible approach, in here referred to as the experimental approach.

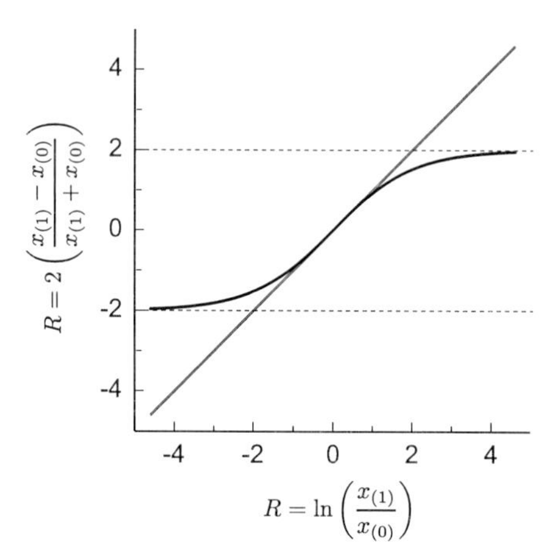

Figure A.7 Comparison of two definitions of the global response Shown is the difference of response coefficients with the experimental response coefficient on the y-axis and the corresponding logarithmic response coefficient on the x-axis (black line). The grey middle line is plotted to visualise the deviation of the two forms. Dashed lines show the asymptotes of the experimental response coefficient function.

In Fig. A.7 the experimental response (y-axis) is compared to the logarithmic response (x-axis). The experimental approach is shown to be capped at ± 2. This is of advantage if large deviations can not be estimated very accurately and might confer an outlier robustness to the experimental approach. The grey line indicates the middle line and shows the deviation of the experimental to the logarithmic response. Since the logarithmic response just denotes log fold changes it can be easily determined that the two forms visibly deviate for R's larger than 1. In other words for fold changes larger than two, the results will differ.

A 6.1 Error propagation

Logarithmic aproach $\mathbf{R}^i_j = \ln\left(\frac{x_{(1)ij}}{x_{(0)i}}\right)$

When neglecting the indices ij and assuming the same CV for $x_{(1)}$ and $x_{(0)}$ Gaussian error propagation of the original error ϵ results in the following error F:

$$
\begin{aligned}
F &= \sqrt{\left(\frac{df(R)}{dx_{(1)}}\epsilon_{(1)}\right)^2 + \left(\frac{df(R)}{dx_{(0)}}\epsilon_{(0)}\right)^2} \\
&= \sqrt{\left(\left(\ln\frac{x_{(1)}}{x_{(0)}}\right)'_{x_{(1)}}\epsilon_{(1)}\right)^2 + \left(\left(\ln\frac{x_{(1)}}{x_{(0)}}\right)'_{x_{(0)}}\epsilon_{(0)}\right)^2} \quad \textit{with} \quad \ln\frac{x_{(1)}}{x_{(0)}} = \ln x_{(1)} - \ln x_{(0)} \\
&= \sqrt{\left(\frac{1}{x_{(1)}}\epsilon_{(1)}\right)^2 + \left(\frac{-1}{x_{(0)}}\epsilon_{(0)}\right)^2} \quad \textit{with} \quad \epsilon_{(1)} = CV \cdot x_{(1)}\ ,\ \epsilon_{(0)} = CV \cdot x_{(0)}
\end{aligned}
$$

$$= \sqrt{2 * CV^2} = \underline{\underline{CV\sqrt{2}}} \quad .$$

The eventual error is therefore independent of the actual size of R.

Experimental approach $\mathbf{R}_j^i = 2\left(\frac{x_{(1)_{ij}} - x_{(0)_i}}{x_{(1)_{ij}} + x_{(0)_i}}\right)$

In this case error propagation results in:

$$F = \sqrt{\left(\frac{df(R)}{dx_{(1)}}\epsilon_{(1)}\right)^2 + \left(\frac{df(R)}{dx_{(0)}}\epsilon_{(0)}\right)^2}$$

$$= \sqrt{\left(\left(\frac{2(x_{(1)} - x_{(0)})}{x_{(1)} + x_{(0)}}\right)'_{x_{(1)}}\epsilon_{(1)}\right)^2 + \left(\left(\frac{2(x_{(1)} - x_{(0)})}{x_{(1)} + x_{(0)}}\right)'_{x_{(0)}}\epsilon_{(0)}\right)^2}$$

$$= \sqrt{\left(\frac{2(x_{(1)} + x_{(0)} - x_{(1)} + x_{(0)})}{(x_{(1)} + x_{(0)})^2}\epsilon_{(1)}\right)^2 + \left(\frac{-2(x_{(1)} + x_{(0)} + x_{(1)} - x_{(0)})}{(x_{(1)} + x_{(0)})^2}\epsilon_{(0)}\right)^2}$$

$$= \sqrt{\left(\frac{4x_{(0)}}{(x_{(1)} + x_{(0)})^2}\epsilon_{(1)}\right)^2 + \left(\frac{-4x_{(1)}}{(x_{(1)} + x_{(0)})^2}\epsilon_{(0)}\right)^2} \quad with \quad \epsilon_{(z)} = CV \cdot x_{(z)},\ z \in \{0, 1\}$$

$$= \sqrt{2CV^2\left(\frac{(4x_{(1)}x_{(0)})^2}{(x_{(1)} + x_{(0)})^4}\right)} \quad with \quad 4x_{(1)}x_{(0)} = (x_{(1)} + x_{(0)})^2 - (x_{(1)} - x_{(0)})^2$$

$$= \sqrt{2CV^2\left(\frac{((x_{(1)} + x_{(0)})^2 - (x_{(1)} - x_{(0)})^2)^2}{(x_{(1)} + x_{(0)})^4}\right)}$$

$$= \sqrt{2CV^2\left(\frac{(x_{(1)} + x_{(0)})^4 + (x_{(1)} - x_{(0)})^4 - 2(x_{(1)} + x_{(0)})^2(x_{(1)} - x_{(0)})^2}{(x_{(1)} + x_{(0)})^4}\right)}$$

$$reformulate\ using \quad R = 2\left(\frac{x_{(1)} - x_{(0)}}{x_{(1)} + x_{(0)}}\right)$$

$$= \sqrt{2CV^2\left(1 + \frac{R^4}{16} - \frac{2R^2}{4}\right)} = \underline{\underline{CV\sqrt{\frac{R^4}{8} - R^2 + 2}}} \quad .$$

Through this derivation the error is dependent on the size of **R**. Moreover, for large **R** the error will shrink and eventually approximate 0. This effect is due to the capping and in practice the error estimation for large R's will thus be underestimated possibly threatening to distort modelling results.

A 6.2 Performance comparison on *in silico* networks

Indeed, by incorporating the two versions of global response coefficient in the ML MS MRA methodology different results can be obtained (Fig. A.8). Especially for transcriptional and posttranslational networks mcc and precision are in general higher for the logarithmic version of $\mathbf{R}$ whereas sensitivity does not favour either side. For the mixed case the improved precision does not translate into a better overall performance (mcc) which is not due to sensitivity. Apparently, this discrepancy stems from the fact that the ML MS MRA based on the logarithmic approach results more often in networks lacking significant links than when based on the capped version. Therefore, although the overall performance of logarithmic $\mathbf{R}$ seems to be superior for most cases, for unfavourable noise scenarios the capped version might be preferable.

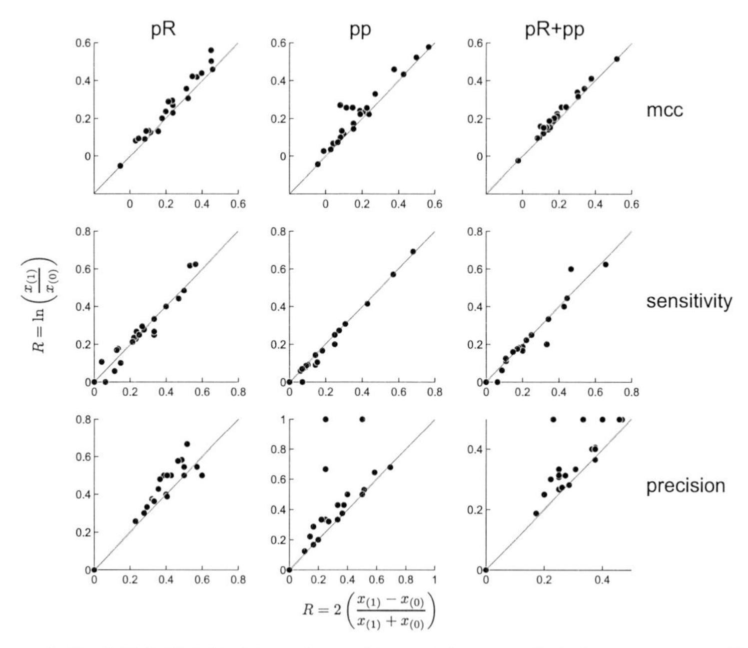

Figure A.8 ML MS MRA benchmarking of the two global response coefficient versions Deviation from parity (grey line) of logarithmic (y-axis) against the capped (x-axis) definition of the global response coefficient. Shown are median statistics for all networks that have been generated for benchmarking of ML MS MRA Suppl. Section A 3 on page 117.

Appendix for chapter: Modelling EGFR signalling in a colon cancer panel

B 1 Phosphorylation kinetics after stimulation

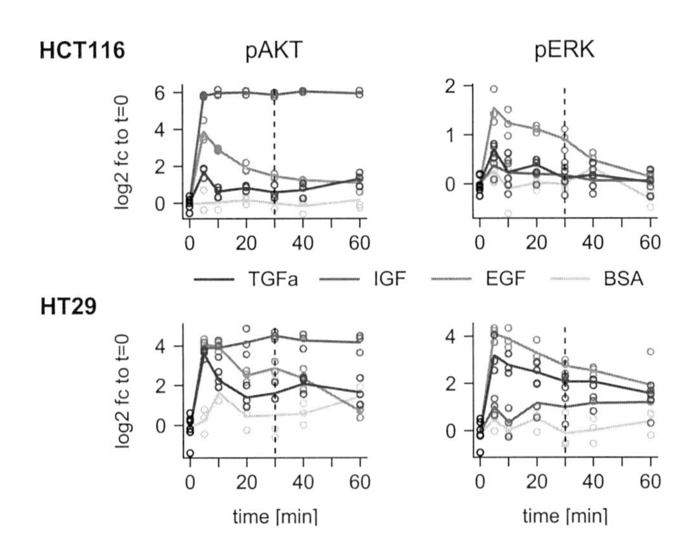

Figure B.1 Phosphorylation kinetics after stimulation Luminex measurements of HCT116 (upper row) and HT29 (lower row) treated for 5-60 min with the indicated ligands demonstrate for the relevant outputs that TGFα(black) as well as IGF (dark grey) signalling is near post-treatment steady state after about 30 min (dashed vertical line). The lines indicate the medium log2 fold change to the average control measurements at t=0 ($n \geq 3$). *Data measured by Anja Sieber*

B 2 Error model for signalling data

For the cell-wise error model replicate measurements for the unperturbed state (4) and replicates for all single inhibitions were used (4 duplicates). In particular, an error model consisting of additive and multiplicative error was used. The multiplicative error was calculated by averaging the Coefficients of Variations (CV) of replicate measurements, and the additive component was taken from background measurements. Subsequently, for each antibody k and each cell line one error model was generated and the error of each measurement j was calculated as:

$$\epsilon_j^k = \text{background}^k + \text{CV}^k \cdot \text{signal}_j^k \tag{B.1}$$

Replicates whose average signal was below twice the background level were not considered for the error model as they do not reliably represent the noise of the biological signal but rather the noise of the measuring device which is usually smaller. If less than three replicate measurements remained, the CV was approximated to 30%. To prevent overfitting, the lower threshold of the CV was set to 10%.

B 3 Model reduction example

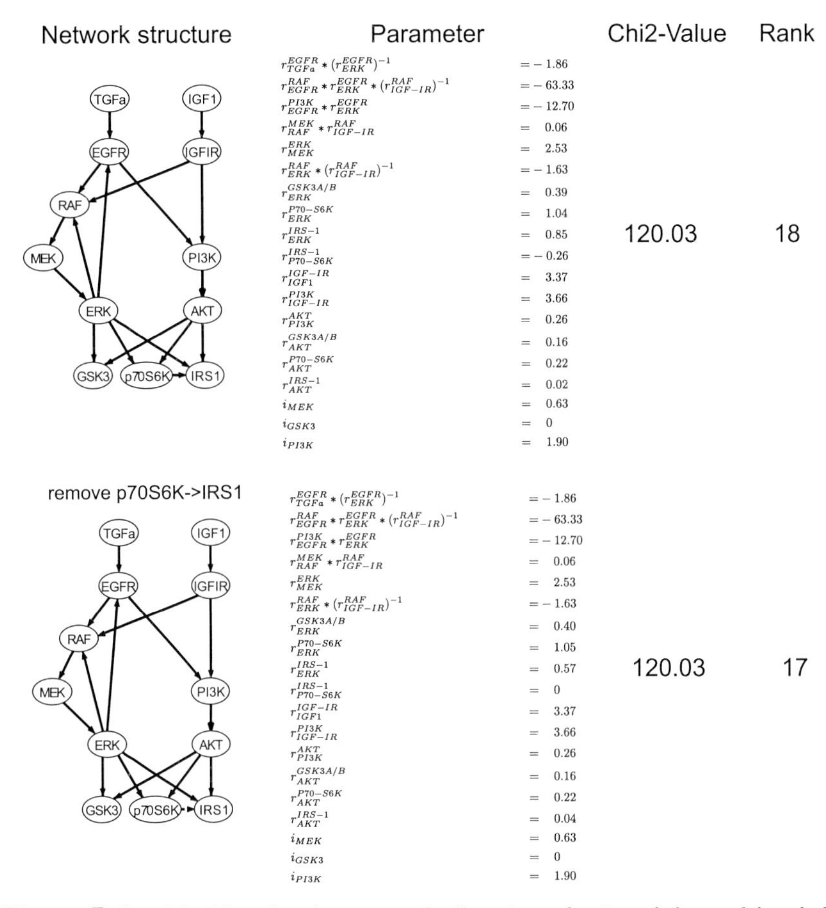

Figure B.2 Model reduction example Stepwise reduction of the model and the corresponding identifiable parameter combinations exemplified on cell line HT29. First column denotes the network structure with dashed lines indicating removed links. Second column depicts the parametrisation (natural logarithm) after fitting the model to the structure. The parameter values of the full model stem from the best fit of 10000 random initialisations. For all further reduced models the best initial fit is given as starting point (except for the removed link which is kept zero). The last two columns show the chi-squared value and the number of identifiable parameters, respectively.

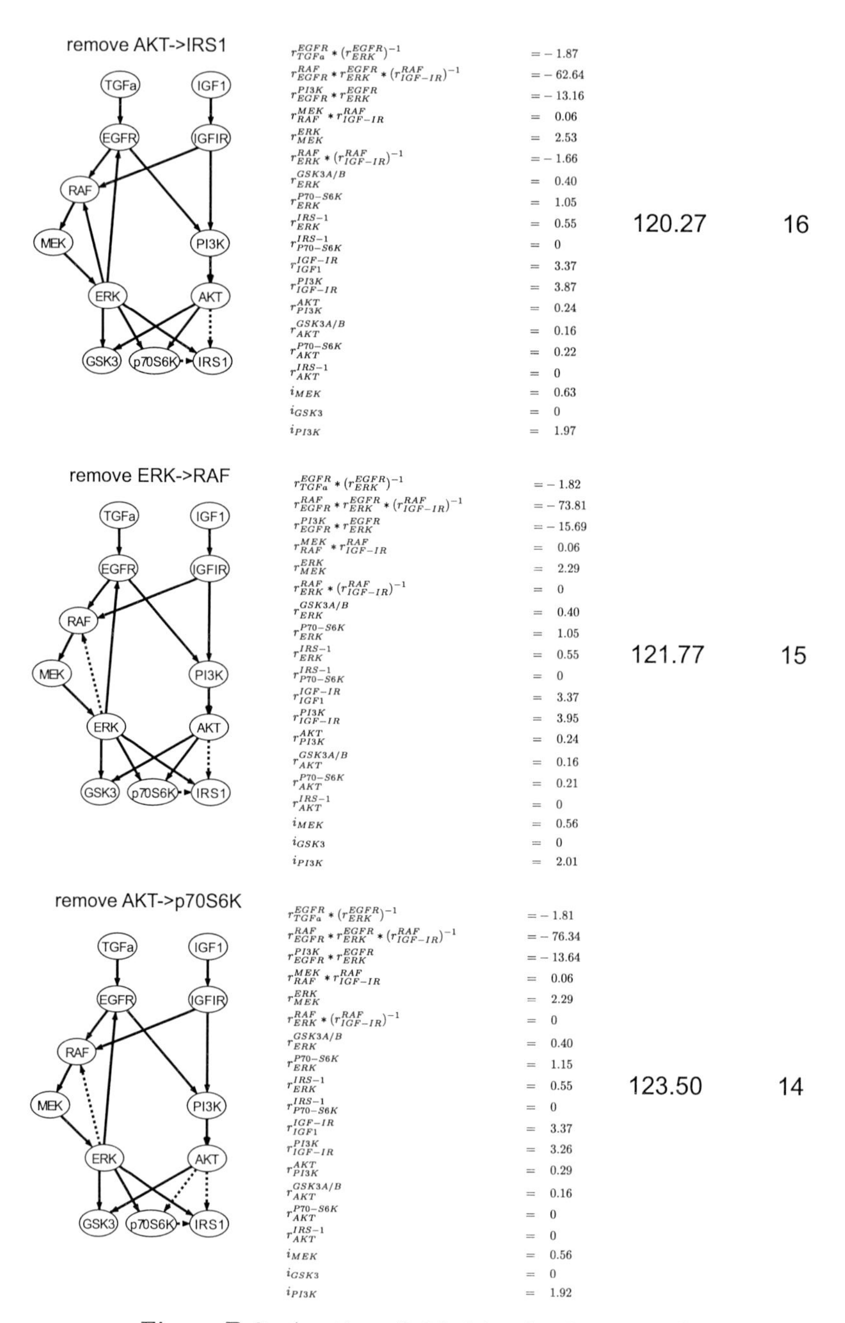

Figure B.2 (continued) Model reduction example

Appendix for chapter: Conceptual modelling resolves role of TTP in HIF-1 regulation

C 1 MRA models to elucidate TTP-HIF-1α relationship

In this section all local response coefficients and global response entries for stimulus-to-ZFP36 and stimulus-to-HIF-target are displayed for each model that was generated in section 4.2.2. Models are nested and newly introduced features will be depicted in bold font in the corresponding local response matrix.

Original Conception

The original model only stated that TTP negatively affects HIF1A (cf. Fig. 4.2). Thus the local response matrix **r** consists of a linear chain from stimulus S to HIF–target.

$$
\begin{array}{l} \\ \text{S} \\ \text{ZFP36} \\ \text{TTP} \\ \text{HIF1A} \\ \text{HIF}-1\alpha \\ \text{HIF}-\text{target} \end{array}
\begin{array}{c}
\begin{array}{cccccc} \text{S} & \text{ZFP36} & \text{TTP} & \text{HIF1A} & \text{HIF}-1\alpha & \text{HIF}-\text{target} \end{array} \\
\left(\begin{array}{cccccc}
-1 & 0 & 0 & 0 & 0 & 0 \\
\mathrm{r}^{\mathrm{S}}_{\mathrm{ZFP36}} & -1 & 0 & 0 & 0 & 0 \\
0 & \mathrm{r}^{\mathrm{ZFP36}}_{\mathrm{TPP}} & -1 & 0 & 0 & 0 \\
0 & 0 & -\mathrm{r}^{\mathrm{TTP}}_{\mathrm{HIF1A}} & -1 & 0 & 0 \\
0 & 0 & 0 & \mathrm{r}^{\mathrm{HIF1A}}_{\mathrm{HIF}-1\alpha} & -1 & 0 \\
0 & 0 & 0 & 0 & \mathrm{r}^{\mathrm{HIF}-1\alpha}_{\mathrm{HIF}-\mathrm{target}} & -1
\end{array}\right)
\end{array}
$$

In this manner the global response entries for $\mathrm{R}^{\mathrm{S}}_{\mathrm{ZFP36}}$ and $\mathrm{R}^{\mathrm{S}}_{\mathrm{HIF-target}}$ will always be anticorrelated :

$$R^{S}_{ZFP36} = \mathrm{r}^{\mathrm{S}}_{\mathrm{ZFP36}} \;,\; R^{S}_{HIF-target} = -\mathrm{r}^{\mathrm{S}}_{\mathrm{ZFP36}}\,\mathrm{r}^{\mathrm{ZFP36}}_{\mathrm{TPP}}\,\mathrm{r}^{\mathrm{TTP}}_{\mathrm{HIF1A}}\,\mathrm{r}^{\mathrm{HIF1A}}_{\mathrm{HIF}-1\alpha}\,\mathrm{r}^{\mathrm{HIF}-1\alpha}_{\mathrm{HIF}-\mathrm{target}} \,. \quad \text{(C.1)}$$

Negative feedback model

This model takes into account that ZFP36 has itself an AU-rich 3'UTR on which its protein product TTP is reported to bind. Thus a negative feedback is introduced into the system, by inserting a negative link between TTP and ZFP36 in the local response matrix.

$$
\begin{array}{c} \\ \text{S} \\ \text{ZFP36} \\ \text{TTP} \\ \text{HIF1A} \\ \text{HIF}-1\alpha \\ \text{HIF}-\text{target} \end{array}
\begin{array}{c} \begin{array}{cccccc} \text{S} & \text{ZFP36} & \text{TTP} & \text{HIF1A} & \text{HIF}-\alpha & \text{HIF}-\text{target} \end{array} \\
\begin{pmatrix}
-1 & 0 & 0 & 0 & 0 & 0 \\
\mathrm{r}^{\mathrm{S}}_{\mathrm{ZFP36}} & -1 & \mathbf{-r^{TTP}_{ZFP36}} & 0 & 0 & 0 \\
0 & \mathrm{r}^{\mathrm{ZFP36}}_{\mathrm{TPP}} & -1 & 0 & 0 & 0 \\
0 & 0 & -\mathrm{r}^{\mathrm{TTP}}_{\mathrm{HIF1A}} & -1 & 0 & 0 \\
0 & 0 & 0 & \mathrm{r}^{\mathrm{HIF1A}}_{\mathrm{HIF}-1\alpha} & -1 & 0 \\
0 & 0 & 0 & 0 & \mathrm{r}^{\mathrm{HIF}-1\alpha}_{\mathrm{HIF}-\mathrm{target}} & -1
\end{pmatrix} \end{array}
$$

Looking at the global response coefficients, the feedback only attenuates the strength but is unable to break the anti-correlation :

$$
R^{S}_{ZFP36} = \frac{\mathrm{r}^{\mathrm{S}}_{\mathrm{ZFP36}}}{\mathrm{r}^{\mathrm{TTP}}_{\mathrm{ZFP36}}\,\mathrm{r}^{\mathrm{ZFP36}}_{\mathrm{TPP}} + 1} \;,\; R^{S}_{HIF-target} = -\frac{\mathrm{r}^{\mathrm{S}}_{\mathrm{ZFP36}}\,\mathrm{r}^{\mathrm{ZFP36}}_{\mathrm{TPP}}\,\mathrm{r}^{\mathrm{TTP}}_{\mathrm{HIF1A}}\,\mathrm{r}^{\mathrm{HIF1A}}_{\mathrm{HIF}-1\alpha}\,\mathrm{r}^{\mathrm{HIF}-1\alpha}_{\mathrm{HIF}-\mathrm{target}}}{\mathrm{r}^{\mathrm{TTP}}_{\mathrm{ZFP36}}\,\mathrm{r}^{\mathrm{ZFP36}}_{\mathrm{TPP}} + 1} \,. \tag{C.2}
$$

Sequestration model

It has been hypothesised that phosphorylation of TTP weakens the binding of TTP to 3'UTR of mRNA and a new node (called pTTP) was introduced into the local response matrix. To model the transformation from active TTP to inactive pTTP, two coefficients were introduced from an unspecified stimulus to both, TTP and pTTP. The two coefficients were coupled in absolute value but of opposite sign.

$$
\begin{array}{c} \\ \text{S} \\ \text{ZFP36} \\ \text{TTP} \\ \mathbf{pTTP} \\ \text{HIF1A} \\ \text{HIF}-1\alpha \\ \text{HIF}-\text{target} \end{array}
\begin{array}{c} \begin{array}{ccccccc} \text{S} & \text{ZFP36} & \text{TTP} & \mathbf{pTTP} & \text{HIF1A} & \text{HIF}-1\alpha & \text{HIF}-\text{target} \end{array} \\
\begin{pmatrix}
-1 & 0 & 0 & 0 & 0 & 0 & 0 \\
\mathrm{r}^{\mathrm{S}}_{\mathrm{ZFP36}} & -1 & -\mathrm{r}^{\mathrm{TTP}}_{\mathrm{ZFP36}} & 0 & 0 & 0 & 0 \\
\mathbf{-r^{S}_{TTP}} & \mathrm{r}^{\mathrm{ZFP36}}_{\mathrm{TPP}} & -1 & 0 & 0 & 0 & 0 \\
\mathbf{r^{S}_{pTTP}} & 0 & 0 & -1 & 0 & 0 & 0 \\
0 & 0 & -\mathrm{r}^{\mathrm{TTP}}_{\mathrm{HIF1A}} & 0 & -1 & 0 & 0 \\
0 & 0 & 0 & 0 & \mathrm{r}^{\mathrm{HIF1A}}_{\mathrm{HIF}-1\alpha} & -1 & 0 \\
0 & 0 & 0 & 0 & 0 & \mathrm{r}^{\mathrm{HIF}-1\alpha}_{\mathrm{HIF}-\mathrm{target}} & -1
\end{pmatrix} \end{array}
$$

It should be mentioned that the stimulus driving TTP phosphorylation does not have to be the same that induces ZFP36 transcription. Therefore stimulus S is standing for the cloud of external environmental cues feeding into the network. For the sequestration model a positive correlation is possible :

$$\begin{aligned} R^{S}_{ZFP36} &= \frac{\mathrm{r}^{\mathrm{S}}_{\mathrm{ZFP36}} + \mathrm{r}^{\mathrm{S}}_{\mathrm{TTP}}\,\mathrm{r}^{\mathrm{TTP}}_{\mathrm{ZFP36}}}{\mathrm{r}^{\mathrm{TTP}}_{\mathrm{ZFP36}}\,\mathrm{r}^{\mathrm{ZFP36}}_{\mathrm{TPP}} + 1} \quad, \\ R^{S}_{HIF-target} &= \frac{\mathrm{r}^{\mathrm{HIF1A}}_{\mathrm{HIF-1}\alpha}\,\mathrm{r}^{\mathrm{HIF-1}\alpha}_{\mathrm{HIF-target}}\,\mathrm{r}^{\mathrm{TTP}}_{\mathrm{HIF1A}}\left(\mathrm{r}^{\mathrm{S}}_{\mathrm{TTP}} - \mathrm{r}^{\mathrm{S}}_{\mathrm{ZFP36}}\,\mathrm{r}^{\mathrm{ZFP36}}_{\mathrm{TPP}}\right)}{\mathrm{r}^{\mathrm{TTP}}_{\mathrm{ZFP36}}\,\mathrm{r}^{\mathrm{ZFP36}}_{\mathrm{TPP}} + 1} \quad. \end{aligned} \tag{C.3}$$

However, the positive correlation only holds for those cases where TTP phosphorylation is higher than TTP production.

Competition models

Other studies postulated a more active role of phosphorylated TTP than sequestration. In order to test several possibilities, two effects were modelled alone and in combinations: (I) Competition for the same binding site with unphosphorylated TTP and (II) active stabilisation of the bound mRNA.

Competitive binding can be realised in different approaches. One realisation would be to allow pTTP to act as an inhibitor of TTP by multiplying an inhibitory term to all outgoing nodes of TTP. The problem then would be that this inhibitory term will only dampen the negative effect but not affect the sign of the regulation and therefore can not produce different results than the sequestration scenario. Another possibility is to assume that free TTP is less stable than bound TTP which can be realised by introducing a negative local response coefficient from pTTP on TTP. Since unphosphorylated TTP is reported to be less stable than phosphorylated TTP (Hitti et al., 2006), this assumption might not be too far fetched. This extension will on one hand cause a reduction of TTP protein and on the other hand an increased TTP production as the negative TTP effect on TTP mRNA is reduced.

$$
\begin{array}{l|ccccccc}
 & \mathrm{S} & \mathrm{ZFP36} & \mathrm{TTP} & \mathrm{pTTP} & \mathrm{HIF1A} & \mathrm{HIF\!-\!1}\alpha & \mathrm{HIF\!-\!target} \\
\mathrm{S} & -1 & 0 & 0 & 0 & 0 & 0 & 0 \\
\mathrm{ZFP36} & \mathrm{r}^{\mathrm{S}}_{\mathrm{ZFP36}} & -1 & -\mathrm{r}^{\mathrm{TTP}}_{\mathrm{ZFP36}} & 0 & 0 & 0 & 0 \\
\mathrm{TTP} & -\mathrm{r}^{\mathrm{S}}_{\mathrm{TTP}} & \mathrm{r}^{\mathrm{ZFP36}}_{\mathrm{TPP}} & -1 & \mathbf{-r^{pTTP}_{TTP}} & 0 & 0 & 0 \\
\mathrm{pTTP} & \mathrm{r}^{\mathrm{S}}_{\mathrm{pTTP}} & 0 & 0 & -1 & 0 & 0 & 0 \\
\mathrm{HIF1A} & 0 & 0 & -\mathrm{r}^{\mathrm{TTP}}_{\mathrm{HIF1A}} & 0 & -1 & 0 & 0 \\
\mathrm{HIF\!-\!1}\alpha & 0 & 0 & 0 & 0 & \mathrm{r}^{\mathrm{HIF1A}}_{\mathrm{HIF-1}\alpha} & -1 & 0 \\
\mathrm{HIF\!-\!target} & 0 & 0 & 0 & 0 & 0 & \mathrm{r}^{\mathrm{HIF-1}\alpha}_{\mathrm{HIF-target}} & -1
\end{array}
$$

The resulting global responses are as follows :

$$
\begin{aligned}
R^S_{ZFP36} &= \frac{\mathrm{r}^{\mathrm{S}}_{\mathrm{ZFP36}} + \mathrm{r}^{\mathrm{S}}_{\mathrm{TTP}}\,\mathrm{r}^{\mathrm{TTP}}_{\mathrm{ZFP36}} + \mathrm{r}^{\mathrm{S}}_{\mathrm{pTTP}}\,\mathrm{r}^{\mathrm{TTP}}_{\mathrm{ZFP36}}\,\mathrm{r}^{\mathrm{pTTP}}_{\mathrm{TTP}}}{\mathrm{r}^{\mathrm{TTP}}_{\mathrm{ZFP36}}\,\mathrm{r}^{\mathrm{ZFP36}}_{\mathrm{TPP}} + 1}, \\
R^S_{HIF-target} &= \frac{\mathrm{r}^{\mathrm{HIF1A}}_{\mathrm{HIF-1}\alpha}\,\mathrm{r}^{\mathrm{HIF-1}\alpha}_{\mathrm{HIF-target}}\,\mathrm{r}^{\mathrm{TTP}}_{\mathrm{HIF1A}}\left(\mathrm{r}^{\mathrm{S}}_{\mathrm{TTP}} - \mathrm{r}^{\mathrm{S}}_{\mathrm{ZFP36}}\,\mathrm{r}^{\mathrm{ZFP36}}_{\mathrm{TPP}} + \mathrm{r}^{\mathrm{S}}_{\mathrm{pTTP}}\,\mathrm{r}^{\mathrm{pTTP}}_{\mathrm{TTP}}\right)}{\mathrm{r}^{\mathrm{TTP}}_{\mathrm{ZFP36}}\,\mathrm{r}^{\mathrm{ZFP36}}_{\mathrm{TPP}} + 1}.
\end{aligned}
\tag{C.4}
$$

The second type, activatory stabilisation, can be realised by adding two positive links from pTTP to the target mRNAs.

$$
\begin{array}{l|ccccccc}
 & \mathrm{S} & \mathrm{ZFP36} & \mathrm{TTP} & \mathrm{pTTP} & \mathrm{HIF1A} & \mathrm{HIF\!-\!1}\alpha & \mathrm{HIF\!-\!target} \\
\mathrm{S} & -1 & 0 & 0 & 0 & 0 & 0 & 0 \\
\mathrm{ZFP36} & \mathrm{r}^{\mathrm{S}}_{\mathrm{ZFP36}} & -1 & -\mathrm{r}^{\mathrm{TTP}}_{\mathrm{ZFP36}} & \mathbf{r^{pTTP}_{ZFP36}} & 0 & 0 & 0 \\
\mathrm{TTP} & -\mathrm{r}^{\mathrm{S}}_{\mathrm{TTP}} & \mathrm{r}^{\mathrm{ZFP36}}_{\mathrm{TPP}} & -1 & 0 & 0 & 0 & 0 \\
\mathrm{pTTP} & \mathrm{r}^{\mathrm{S}}_{\mathrm{pTTP}} & 0 & 0 & -1 & 0 & 0 & 0 \\
\mathrm{HIF1A} & 0 & 0 & -\mathrm{r}^{\mathrm{TTP}}_{\mathrm{HIF1A}} & \mathbf{r^{pTTP}_{ZFP36}} & -1 & 0 & 0 \\
\mathrm{HIF\!-\!1}\alpha & 0 & 0 & 0 & 0 & \mathrm{r}^{\mathrm{HIF1A}}_{\mathrm{HIF-1}\alpha} & -1 & 0 \\
\mathrm{HIF\!-\!target} & 0 & 0 & 0 & 0 & 0 & \mathrm{r}^{\mathrm{HIF-1}\alpha}_{\mathrm{HIF-target}} & -1
\end{array}
$$

Whereas the global response coefficient from the stimulus to ZFP36 shows a constant sign, the global response of the HIF-target can vary in a non-trivial fashion:

$$
\begin{aligned}
R^S_{ZFP36} &= \frac{\mathrm{r}^{\mathrm{S}}_{\mathrm{ZFP36}} + \mathrm{r}^{\mathrm{S}}_{\mathrm{TTP}}\,\mathrm{r}^{\mathrm{TTP}}_{\mathrm{ZFP36}} + \mathrm{r}^{\mathrm{S}}_{\mathrm{pTTP}}\,\mathrm{r}^{\mathrm{pTTP}}_{\mathrm{ZFP36}}}{\mathrm{r}^{\mathrm{TTP}}_{\mathrm{ZFP36}}\,\mathrm{r}^{\mathrm{ZFP36}}_{\mathrm{TPP}} + 1}, \\
R^S_{HIF-target} &= \mathrm{r}^{\mathrm{HIF1A}}_{\mathrm{HIF-1}\alpha}\,\mathrm{r}^{\mathrm{HIF-1}\alpha}_{\mathrm{HIF-target}}\left(\mathrm{r}^{\mathrm{S}}_{\mathrm{pTTP}}\,\mathrm{r}^{\mathrm{pTTP}}_{\mathrm{HIF1A}} - \right. \\
&\left. \frac{\mathrm{r}^{\mathrm{TTP}}_{\mathrm{HIF1A}}\left(\mathrm{r}^{\mathrm{S}}_{\mathrm{ZFP36}}\,\mathrm{r}^{\mathrm{ZFP36}}_{\mathrm{TPP}} + \mathrm{r}^{\mathrm{S}}_{\mathrm{pTTP}}\,\mathrm{r}^{\mathrm{ZFP36}}_{\mathrm{TPP}}\,\mathrm{r}^{\mathrm{pTTP}}_{\mathrm{ZFP36}} - \mathrm{r}^{\mathrm{S}}_{\mathrm{TTP}}\right)}{\mathrm{r}^{\mathrm{TTP}}_{\mathrm{ZFP36}}\,\mathrm{r}^{\mathrm{ZFP36}}_{\mathrm{TPP}} + 1}\right).
\end{aligned}
\tag{C.5}
$$

Combined effects As both effects are not mutually exclusive in the regulatory sense as well as in their representation in the local response matrix, the effect of both mechanisms was modelled. Simulations showed an even higher positive correlation which can not be deduced by simply studying the global response terms:

$$
\begin{aligned}
R^S_{ZFP36} &= \frac{\mathrm{r}^{\mathrm{S}}_{\mathrm{ZFP36}} + \mathrm{r}^{\mathrm{S}}_{\mathrm{TTP}}\,\mathrm{r}^{\mathrm{TTP}}_{\mathrm{ZFP36}} + \mathrm{r}^{\mathrm{S}}_{\mathrm{pTTP}}\,\mathrm{r}^{\mathrm{pTTP}}_{\mathrm{ZFP36}} + \mathrm{r}^{\mathrm{S}}_{\mathrm{pTTP}}\,\mathrm{r}^{\mathrm{TTP}}_{\mathrm{ZFP36}}\,\mathrm{r}^{\mathrm{pTTP}}_{\mathrm{TTP}}}{\mathrm{r}^{\mathrm{TTP}}_{\mathrm{ZFP36}}\,\mathrm{r}^{\mathrm{ZFP36}}_{\mathrm{TPP}} + 1}, \\
R^S_{HIF-target} &= \mathrm{r}^{\mathrm{HIF1A}}_{\mathrm{HIF-1\alpha}}\,\mathrm{r}^{\mathrm{HIF-1\alpha}}_{\mathrm{HIF-target}}\Bigg(\mathrm{r}^{\mathrm{S}}_{\mathrm{pTTP}}\,\mathrm{r}^{\mathrm{pTTP}}_{\mathrm{HIF1A}} + \\
&\quad \frac{\mathrm{r}^{\mathrm{TTP}}_{\mathrm{HIF1A}}\left(\mathrm{r}^{\mathrm{S}}_{\mathrm{TTP}} - \mathrm{r}^{\mathrm{S}}_{\mathrm{ZFP36}}\,\mathrm{r}^{\mathrm{ZFP36}}_{\mathrm{TPP}} + \mathrm{r}^{\mathrm{S}}_{\mathrm{pTTP}}\left(\mathrm{r}^{\mathrm{pTTP}}_{\mathrm{TTP}} - \mathrm{r}^{\mathrm{ZFP36}}_{\mathrm{TPP}}\,\mathrm{r}^{\mathrm{pTTP}}_{\mathrm{ZFP36}}\right)\right)}{\mathrm{r}^{\mathrm{TTP}}_{\mathrm{ZFP36}}\,\mathrm{r}^{\mathrm{ZFP36}}_{\mathrm{TPP}} + 1}\Bigg).
\end{aligned}
\tag{C.6}
$$

Afternote

Next to the investigated correlation pair an alternative choice to derive correlative behaviour exists. One should be able to observe the same results by correlating the sign of the global response coefficient from ZFP36 to HIF-target ($\mathrm{R}^{\mathrm{ZFP36}}_{\mathrm{HIF-target}}$) with the global self-perturbation entry of ZFP36 ($\mathrm{R}^{\mathrm{ZFP36}}_{\mathrm{ZFP36}}$).

Appendix for chapter: Conclusion

D 1 FDA approved targeted therapy drugs

Table D.1 FDA approved targeted therapies Clinically admitted targeted therapies in the USA as of March 2015 ordered by first admission year. In addition to the approval year also drug type, treatable cancers and effect is listed for all drugs. If the drug was interfering with cancer signalling, the level of interference was indicated in the last column. Information was retrieved from the national cancer institute `http://www.cancer.gov/cancertopics/factsheet/Therapy/targeted` the Blue Ridge Institute `http://www.brimr.org/PKI/PKIs.htm` and information of additional targets was derived from pubchem `http://pubchem.ncbi.nlm.nih.gov/`. Note that drugs targeting general mechanisms such as proteasome and microtubule depolymerisation were not included due to their similarity to chemotherapeutic approaches.

Name	FDA app.	Type	Target	Cancer type implication	Effect	Interruption level
anastrozole	1995	aromatase inhibitor	aromatase	Breast cancer	signal interrupt	signal creation
toremifene	1997	ligand antagonist	estrogen receptor	Breast cancer	signal interrupt	reception
tamoxifen	1998	ligand antagonist	estrogen receptor	Breast cancer	signal interrupt	reception
imatinib mesylate	2001	kinase inhibitor	Bcr-Abl, PDGFR, Kit	Dermatofibrosarcoma protuberans, Gastrointestinal stromal tumour, Leukemia, Myelodysplastic proliferative disorders, Systemic mastocytosis	signal interrupt	signal creation, reception
alemtuzumab	2001	antibody	CD52	Leukemia	immunotherapy	
fulvestrant	2002	ligand antagonist	estrogen receptor	Breast cancer	signal interrupt	reception
Ibritumomab tiuxetan	2002	antibody	CD20	Lymphoma	immunotherapy	
gefitinib	2003	kinase inhibitor	EGFR	Lung cancer	signal interrupt	reception
letrozole	2004	aromatase inhibitor	aromatase	Breast cancer	signal interrupt	signal creation
bevacizumab	2004	antibody	VEGF	Brain cancer, Cervical cancer, Colorectal cancer, Kidney cancer, Lung cancer, Ovarian cancer	signal interrupt	reception
erlotinib	2004	kinase inhibitor	EGFR	Lung cancer, Pancreatic cancer	signal interrupt	reception
cetuximab	2004	antibody	EGFR	Colorectal cancer, Head and neck cancer	signal interrupt, immunotherapy	reception
exemestane	2005	aromatase inhibitor	aromatase	Breast cancer	signal interrupt	signal creation
sorafenib	2005	kinase inhibitor	RAF, VEGFR, PDGFR, Kit, RET, FGFR, FLT3	Kidney cancer, Liver cancer, Thyroid cancer	signal interrupt	reception, downstream relay
vorinostat	2006	deacetylase inhibitor	HDAC	Lymphoma	signal interrupt	downstream relay

sunitinib	2006	kinase inhibitor	VEGFR, PDGFR, Kit, FLT3, RET, CSF-1R	Gastrointestinal stromal tumour, Kidney cancer, Pancreatic cancer	signal interrupt	signal creation, reception
dasatinib	2006	kinase inhibitor	Bcr-Abl, Src, STAT5B, PDGFR, Kit, YES1, Lck, EPHA2	Leukemia	signal interrupt	signal creation, reception, downstream relay
panitumumab	2006	antibody	EGFR	Colorectal cancer	signal interrupt, immunotherapy	reception
rituximab	2006	antibody	CD20	Leukemia, Lymphoma	immunotherapy	
temsirolimus	2007	kinase inhibitor	mTOR	Kidney cancer	signal interrupt	downstream relay
lapatinib	2007	kinase inhibitor	EGFR, HER-2	Breast cancer	signal interrupt	signal creation, reception
nilotinib	2007	kinase inhibitor	Bcr-Abl, Kit, PDGFR	Leukemia	signal interrupt	signal creation, reception
denileukin diftitox	2008	ligand agonist	IL2RG	Lymphoma	signal interrupt	reception
everolimus	2009	kinase inhibitor	mTOR, FKBP12	Brain cancer, Breast cancer, Kidney cancer, Pancreatic cancer	signal interrupt	downstream relay
romidepsin	2009	deacetylase inhibitor	HDAC	Lymphoma	signal interrupt	downstream relay
pazopanib	2009	kinase inhibitor	VEGFR, PDGFR, FGFR, Kit, Lck, Fms, Itk	Kidney cancer, Soft tissue sarcoma	signal interrupt	reception, downstream relay
ofatumumab	2009	antibody	CD20	Leukemia	immunotherapy	
pralatrexate	2009	dihydrofolate reductase inhibitor	DHFR	Lymphoma	targeted cell death inducer	
denosumab	2010	antibody	RANKL	Giant cell tumour of the bone	prevents osteolysis but does not target malignant cells	
ruxolitinib phosphate	2011	kinase inhibitor	JAK, TYK2	Myelodysplastic proliferative disorders	signal interrupt	downstream relay
vemurafenib	2011	kinase inhibitor	BRAF (V600E)	Melanoma	signal interrupt	signal creation
crizotinib	2011	kinase inhibitor	c-Met, ALK, ROS1	Lung cancer	signal interrupt	signal creation, reception
vandetanib	2011	kinase inhibitor	VEGFR-2, EGFR, RET, Brk, Tie2, EphR, Src	Thyroid cancer	signal interrupt	signal creation, reception, downstream relay
ipilimumab	2011	antibody	CTLA4	Melanoma	immunotherapy	
brentuximab vedotin	2011	antibody	CD30	Lymphoma	targeted cell death inducer	
pertuzumab	2012	antibody	HER-2	Breast cancer	signal interrupt, immunotherapy	signal creation
bosutinib	2012	kinase inhibitor	Bcr-Abl, BCR, Lyn, HCK, Src, MEK, CDK2, MAP3K2	Leukemia	signal interrupt	signal creation, downstream relay
cabozantinib-s-malate	2012	kinase inhibitor	c-Met, RET, VEGFR, KIT, FLT-3, TIE-2, TRKB, AXL	Thyroid cancer	signal interrupt	signal creation, reception
ponatinib	2012	kinase inhibitor	Bcr-Abl, VEGFR, PDGFR, FGFR, Eph, Src, Kit, RET, Tie2, Flt-3	Leukemia	signal interrupt	signal creation, reception, downstream relay
regorafenib	2012	kinase inhibitor	VEGFR, Ret, Kit, PDGFR, Raf, FGFR, DDR2, TrkA, Eph2A, SAPK2, PTK5, Abl	Colorectal cancer, Gastrointestinal stromal tumour	signal interrupt	signal creation, reception, downstream relay
axitinib	2012	kinase inhibitor	VEGFR, PDGFR	Kidney cancer	signal interrupt	reception
enzalutamide	2012	ligand antagonist	androgene receptor	Prostate cancer	signal interrupt	reception
vismodegib	2012	ligand antagonist	PTCH, SMO	Basal cell carcinoma	signal interrupt	reception
ziv-aflibercept	2012	ligand trap	VEGF	Colorectal cancer	signal interrupt	reception
ibrutinib	2013	kinase inhibitor	BTK	Leukemia, Lymphoma	signal interrupt	downstream relay

trametinib	2013	kinase inhibitor	MEK	Melanoma with BRAF V600E mutation	signal interrupt	downstream relay
Ado-trastuzumab Emtansine	2013	antibody	HER-2	adenocarcinoma of stomach and gastroesophageal junction, Breast cancer	signal interrupt, immunotherapy	signal creation
dabrafenib	2013	kinase inhibitor	RAF, SIK1, NEK11, LIMK1	Melanoma	signal interrupt	signal creation, downstream relay
afatinib dimaleate	2013	kinase inhibitor	EGFR, HER-2, HER-4	Lung cancer	signal interrupt	signal creation, reception
obinutuzumab	2013	antibody	CD20	Leukemia	immunotherapy	
idelalisib	2014	kinase inhibitor	PI3K-delta	Leukemia, Lymphoma	signal interrupt	downstream relay
belinostat	2014	deacetylase inhibitor	HDAC	Lymphoma	signal interrupt	downstream relay
ceritinib	2014	kinase inhibitor	ALK, IGF-1R, InsR, ROS1	Lung cancer	signal interrupt	signal creation, reception
lanreotide acetate	2014	ligand antagonist	SSTR2, SSTR5	(neuro)endocrine tumors	signal interrupt	reception
ramucirumab	2014	antibody	VEGFR-2	adenocarcinoma of stomach and gastroesophageal junction, Lung cancer	signal interrupt	reception
siltuximab	2014	antibody	IL-6	Lymphoma	signal interrupt	reception
blinatumomab	2014	antibody	CD19, CD3	Leukemia	immunotherapy	
nivolumab	2014	antibody	PCD-1	Melanoma	immunotherapy	
olaparib	2014	polymerase inhibitor	PARP	Ovarian cancer with defective BRCA	synthetic lethality	
pembrolizumab	2014	antibody	PCD-1	Melanoma	immunotherapy	
palbociclib	2015	kinase inhibitor	CDK4, 6	Breast cancer	signal interrupt	downstream relay
lenvatinib	2015	kinase inhibitor	VEGFR, FGFR, PDGFR, Kit, RET	Thyroid cancer	signal interrupt	reception

Abbreviations

Abl	ABL proto-oncogene 1, non-receptor tyrosine kinase
AKT	v-akt murine thymoma viral oncogene homolog, isoforms AKT1,2 and 3
ARE	AU-rich elements
ARE-BP	ARE-binding protein
AUROC	area under the receiver operation curve
AUPR	area under the precision recall curve
Bcr	breakpoint cluster region
BVSA	Bayesian variance selection analysis
c-Fos	FBJ murine osteosarcoma viral oncogene homolog, a transcription factor
c-Jun	v-jun avian sarcoma virus 17 oncogene homolog, a transcription factor
ChIP	chromatin immunoprecipitation
CRISPR	clustered regularly interspaced short palindromic repeats
DMSO	dimethyl sulfoxide, solvent control
DREAM	dialogue for reverse engineering assessments and methods
DUSP	dual-specificity phosphatase
EGF	epidermal growth factor
EGFR	EGF receptor, also known as ErbB1 (erythroblastic leukaemia viral oncogene homolog 1)
EGR1	early growth response 1, a transcription factor
ELK1	ETS-like gene 1
ENCODE	encyclopedia of DNA elements
ERK	extracellular signal-regulated kinase
Fosl1	FOS-Like Antigen 1
GEO	gene expression omnibus, gene expression data base
Grb2	growth factor receptor-bound protein 2, an adaptor protein
GSK3A/B	glycogen synthase kinase 3 alpha and beta
HCT116	a colon cancer-derived cell line

HIF-1	hypoxia induced factor 1, consist of HIF-1α and HIF-1β
HNRPA2B1	heterogeneous nuclear ribonucleoproteins A2/B1
HSPA8	heat shock 70kDa protein 8
HT29	a colon cancer derived cell line
IFG	insulin-like growth factor
IGF-IR	insulin-like growth factor 1 receptor
IkB-α	nuclear factor kappa-B kinase
IKK	inhibitor of nuclear factor kappa-B kinase
IRS-1	insulin receptor substrate 1
JAK	janus kinase
JNK	c-Jun N-terminal kinase
Lim1215	a colon cancer-derived cell line
MAPK	mitogen activated protein kinase (synonym to ERK)
MC MRA	Monte Carlo sampling-based MRA
MCMC	Markov-Chain-Monte-Carlo
MEK	kinase of ERK
MK2	MAP Kinase-Activated Protein Kinase 2
MLMSMRA	maximum likelihood model selection MRA
MRA	modular response analysis
Myc	v-myc avian myelocytomatosis viral oncogene homolog, a transcription factor
NFκB	nuclear factor of kappa B
NIR	network identification by multiple regression
ODE	ordinary differential equation
Otx-1	Orthodenticle Homeobox 1, transcription factor
p38	p38 mitogen activated protein kinase
p53	tumor protein p53
PI3K	phosphatidylinositol-4-phosphate 3 kinase
PMA	phorbol-12-myristate-12-acetate
p90RSK	AGC kinase of the RSK family
p70S6K	AGC kinase of the RSK family
pR	protein-RNA interaction
pp	protein-protein interaction
qPCR	quantitative PCR (polymerase chain reaction)
RAF	v-raf-1 murine leukemia viral oncogene, a MAPKK kinase, isoforms A-Raf, B-raf and c-Raf (RAF1)

RAS	rat sarcoma viral oncogene, a GTPase, isoforms NRAS, KRAS and HRAS
RKO	a colon cancer derived cell line
RNA	Ribonucleic acid in here synonymous with mRNA
ROSE	rat ovarian surface epithelium
SDE	stochastic differential equation
siRNA	small interfering RNA
SMAD	similar to mothers against decapentaplegic
SOS	son of sevenless, a guanine nucleotide exchange factor, isoforms Sos1,2
SPEED	signaling pathway enrichment using experimental datasets
STAT	signal transducer and activator of transcription
SW480	a colon cancer derived cell line
TAZ	taffazin
TF	transcription factor
TGFβ	transforming growth factor beta
TGI	tumour growth index
TLR	toll-like receptor
TNFα	tumor necrosis factor alpha
ut	untreated
wt	wild type

Bibliography

Adams, G. P. and Weiner, L. M. 2005. Monoclonal antibody therapy of cancer. Nat Biotechnol, 23(9):1147–57.

Aksamitiene, E., Kholodenko, B. N., Kolch, W., Hoek, J. B., and Kiyatkin, A. 2010. PI3K/Akt-sensitive MEK-independent compensatory circuit of ERK activation in ER-positive PI3K-mutant T47D breast cancer cells. Cell Signal, 22(9):1369–78.

Amado, R. G., Wolf, M., Peeters, M., Cutsem, E. V., Siena, S., Freeman, D. J., Juan, T., Sikorski, R., Suggs, S., Radinsky, R., Patterson, S. D., and Chang, D. D. 2008. Wild-type KRAS is required for panitumumab efficacy in patients with metastatic colorectal cancer. J Clin Oncol, 26(10):1626–34.

Amit, I., Citri, A., Shay, T., Lu, Y., Katz, M., Zhang, F., Tarcic, G., Siwak, D., Lahad, J., Jacob-Hirsch, J., Amariglio, N., Vaisman, N., Segal, E., Rechavi, G., Alon, U., Mills, G. B., Domany, E., and Yarden, Y. 2007. A module of negative feedback regulators defines growth factor signaling. Nat Genet, 39(4):503–12.

Anderson, N. G., Li, P., Marsden, L. A., Williams, N., Roberts, T. M., and Sturgill, T. W. 1991. Raf-1 is a potential substrate for mitogen-activated protein kinase in vivo. Biochem J, 277 (Pt 2):573–6.

Andrec, M., Kholodenko, B. N., Levy, R. M., and Sontag, E. 2005. Inference of signaling and gene regulatory networks by steady-state perturbation experiments: structure and accuracy. J Theor Biol, 232(3):427–41.

Ashburner, M., Ball, C. A., Blake, J. A., Botstein, D., Butler, H., Cherry, J. M., Davis, A. P., Dolinski, K., Dwight, S. S., Eppig, J. T., Harris, M. A., Hill, D. P., Issel-Tarver, L., Kasarskis, A., Lewis, S., Matese, J. C., Richardson, J. E., Ringwald, M., Rubin, G. M., and Sherlock, G. 2000. Gene ontology: tool for the unification of biology. The Gene Ontology Consortium. Nat Genet, 25(1):25–9.

Avraham, R. and Yarden, Y. 2011. Feedback regulation of EGFR signalling: decision making by early and delayed loops. Nat Rev Mol Cell Biol, 12(2): 104–17.

Azzolin, L., Zanconato, F., Bresolin, S., Forcato, M., Basso, G., Bicciato, S., Cordenonsi, M., and Piccolo, S. 2012. Role of TAZ as Mediator of Wnt Signaling. Cell, 151(7):1443–56.

Baines, A. T., Xu, D., and Der, C. J. 2011. Inhibition of Ras for cancer treatment: the search continues. Future Med Chem, 3(14):1787–808.

Balan, V., Leicht, D. T., Zhu, J., Balan, K., Kaplun, A., Singh-Gupta, V., Qin, J., Ruan, H., Comb, M. J., and Tzivion, G. 2006. Identification of novel in vivo Raf-1 phosphorylation sites mediating positive feedback Raf-1 regulation by extracellular signal-regulated kinase. Mol Biol Cell, 17(3):1141–53.

Baldi, P., Brunak, S., Chauvin, Y., Andersen, C. A., and Nielsen, H. 2000. Assessing the accuracy of prediction algorithms for classification: an overview. Bioinformatics, 16(5):412–24.

Balmanno, K., Chell, S. D., Gillings, A. S., Hayat, S., and Cook, S. J. 2009. Intrinsic resistance to the MEK1/2 inhibitor AZD6244 (ARRY-142886) is associated with weak ERK1/2 signalling and/or strong PI3K signalling in colorectal cancer cell lines. Int J Cancer, 125(10):2332–41.

Barenco, M., Tomescu, D., Brewer, D., Callard, R., Stark, J., and Hubank, M. 2006. Ranked prediction of p53 targets using hidden variable dynamic modeling. Genome Biol, 7(3):R25.

Barrett, T., Wilhite, S. E., Ledoux, P., Evangelista, C., Kim, I. F., Tomashevsky, M., Marshall, K. A., Phillippy, K. H., Sherman, P. M., Holko, M., Yefanov, A., Lee, H., Zhang, N., Robertson, C. L., Serova, N., Davis, S., and Soboleva, A. 2013. NCBI GEO: archive for functional genomics data sets–update. Nucleic Acids Res, 41(Database issue):D991–5.

Bermudez, O., Jouandin, P., Rottier, J., Bourcier, C., Pagès, G., and Gimond, C. 2011. Post-transcriptional regulation of the DUSP6/MKP-3 phosphatase by MEK/ERK signaling and hypoxia. J. Cell. Physiol., 226(1):276–84.

Birtwistle, M. R., Hatakeyama, M., Yumoto, N., Ogunnaike, B. A., Hoek, J. B., and Kholodenko, B. N. 2007. Ligand-dependent responses of the ErbB signaling network: experimental and modeling analyses. Molecular Systems Biology, 3: 144.

Blackshear, P. J. 2002. Tristetraprolin and other CCCH tandem zinc-finger proteins in the regulation of mRNA turnover. Biochem Soc Trans, 30(Pt 6):945–52.

Blüthgen, N. 2010. Transcriptional feedbacks in mammalian signal transduction pathways facilitate rapid and reliable protein induction. Mol Biosyst, 6(7):1277–84.

Blüthgen, N. and Legewie, S. 2008. Systems analysis of MAPK signal transduction. Essays Biochem, 45:95–107.

Blüthgen, N., Legewie, S., Kielbasa, S. M., Schramme, A., Tchernitsa, O., Keil, J., Solf, A., Vingron, M., Schäfer, R., Herzel, H., and Sers, C. 2009. A systems biological approach suggests that transcriptional feedback regulation by dual-specificity phosphatase 6 shapes extracellular signal-related kinase activity in RAS-transformed fibroblasts. FEBS J, 276(4):1024–35.

Bolouri, H. 2014. Modeling genomic regulatory networks with big data. Trends Genet, 30(5):182–91.

Bovolenta, L. A., Acencio, M. L., and Lemke, N. 2012. HTRIdb: an open-access database for experimentally verified human transcriptional regulation interactions. BMC genomics, 13(1):405.

Box, G. E. P. Robustness in the strategy of scientific model building. In Robustness in statistics, pages 201–36. Elsevier, 1979. ISBN 978-0-12-438150-6.

Brahma, P. K., Zhang, H., Murray, B. S., jue Shu, F., Sidell, N., Seli, E., and Kallen, C. B. 2012. The mRNA-binding protein Zfp36 is upregulated by β-adrenergic stimulation and represses IL-6 production in 3T3-L1 adipocytes. Obesity (Silver Spring), 20(1):40–7.

Brazhnik, P. 2005. Inferring gene networks from steady-state response to single-gene perturbations. J Theor Biol, 237(4):427–40.

Britschgi, A., Andraos, R., Brinkhaus, H., Klebba, I., Romanet, V., Müller, U., Murakami, M., Radimerski, T., and Bentires-Alj, M. 2012. JAK2/STAT5 inhibition circumvents resistance to PI3K/mTOR blockade: a rationale for cotargeting these pathways in metastatic breast cancer. Cancer Cell, 22(6):796–811.

Brooks, S. A. and Blackshear, P. J. 2013. Tristetraprolin (TTP): interactions with mRNA and proteins, and current thoughts on mechanisms of action. Biochim Biophys Acta, 1829(6-7):666–79.

Brooks, S. A., Connolly, J. E., Diegel, R. J., Fava, R. A., and Rigby, W. F. C. 2002. Analysis of the function, expression, and subcellular distribution of human tristetraprolin. Arthritis Rheum, 46(5):1362–70.

Brooks, S. A., Connolly, J. E., and Rigby, W. F. C. 2004. The role of mRNA turnover in the regulation of tristetraprolin expression: evidence for an extracellular signal-regulated kinase-specific, AU-rich element-dependent, autoregulatory pathway. J Immunol, 172(12):7263–71.

Brough, R., Frankum, J. R., Costa-Cabral, S., Lord, C. J., and Ashworth, A. 2011. Searching for synthetic lethality in cancer. Curr Opin Genet Dev, 21(1):34–41.

Bruggeman, F. J., Westerhoff, H. V., Hoek, J. B., and Kholodenko, B. N. 2002. Modular response analysis of cellular regulatory networks. J Theor Biol, 218(4): 507–20.

Bublil, E. M. and Yarden, Y. 2007. The EGF receptor family: spearheading a merger of signaling and therapeutics. Curr Opin Cell Biol, 19(2):124–34.

Burack, W. R. and Shaw, A. S. 2005. Live Cell Imaging of ERK and MEK: simple binding equilibrium explains the regulated nucleocytoplasmic distribution of ERK. J Biol Chem, 280(5):3832–7.

Burke, J. R., Pattoli, M. A., Gregor, K. R., Brassil, P. J., MacMaster, J. F., McIntyre, K. W., Yang, X., Iotzova, V. S., Clarke, W., Strnad, J., Qiu, Y., and Zusi, F. C. 2003. BMS-345541 is a highly selective inhibitor of I kappa B kinase that binds at an allosteric site of the enzyme and blocks NF-kappa B-dependent transcription in mice. J Biol Chem, 278(3):1450–6.

Cairns, R. A., Harris, I. S., and Mak, T. W. 2011. Regulation of cancer cell metabolism. Nat Rev Cancer, 11(2):85–95.

Camacho, D., Licona, P. V., Mendes, P., and Laubenbacher, R. 2007. Comparison of reverse-engineering methods using an in silico network. Ann N Y Acad Sci, 1115:73–89.

Cantone, I., Marucci, L., Iorio, F., Ricci, M. A., Belcastro, V., Bansal, M., Santini, S., di Bernardo, M., di Bernardo, D., and Cosma, M. P. 2009. A yeast synthetic network for in vivo assessment of reverse-engineering and modeling approaches. Cell, 137(1):172–81.

Cao, H., Tuttle, J. S., and Blackshear, P. J. 2004. Immunological characterization of tristetraprolin as a low abundance, inducible, stable cytosolic protein. J Biol Chem, 279(20):21489–99.

Cao, H., Deterding, L. J., Venable, J. D., Kennington, E. A., Yates, J. R., Tomer, K. B., and Blackshear, P. J. 2006. Identification of the anti-inflammatory protein tristetraprolin as a hyperphosphorylated protein by mass spectrometry and site-directed mutagenesis. Biochem J, 394(Pt 1):285–97.

Carballo, E., Cao, H., Lai, W. S., Kennington, E. A., Campbell, D., and Blackshear, P. J. 2001. Decreased sensitivity of tristetraprolin-deficient cells to p38 inhibitors suggests the involvement of tristetraprolin in the p38 signaling pathway. J Biol Chem, 276(45):42580–7.

Casalino, L., Cesare, D. D., and Verde, P. 2003. Accumulation of Fra-1 in ras-transformed cells depends on both transcriptional autoregulation and MEK-dependent posttranslational stabilization. Mol Cell Biol, 23(12):4401–15.

Cavadas, M. A., Nguyen, L. K., and Cheong, A. 2013. Hypoxia-inducible factor (HIF) network: insights from mathematical models. Cell Commun Signal, 11(1): 42.

Cedersund, G., Roll, J., Ulfhielm, E., Danielsson, A., Tidefelt, H., and Strålfors, P. 2008. Model-based hypothesis testing of key mechanisms in initial phase of insulin signaling. PLoS Comput Biol, 4(6):e1000096.

Chamboredon, S., Ciais, D., Desroches-Castan, A., Savi, P., Bono, F., Feige, J.-J., and Cherradi, N. 2011. Hypoxia-inducible factor-1α mRNA: a new target for destabilization by tristetraprolin in endothelial cells. Mol Biol Cell, 22(18): 3366–78.

Chandarlapaty, S., Sawai, A., Scaltriti, M., Rodrik-Outmezguine, V., Grbovic-Huezo, O., Serra, V., Majumder, P. K., Baselga, J., and Rosen, N. 2011. AKT inhibition relieves feedback suppression of receptor tyrosine kinase expression and activity. Cancer Cell, 19(1):58–71.

Chen, Y.-L., Jiang, Y.-W., Su, Y.-L., Lee, S.-C., Chang, M.-S., and Chang, C.-J. 2013. Transcriptional regulation of tristetraprolin by NF-κB signaling in LPS-stimulated macrophages. Mol Biol Rep, 40(4):2867–77.

Cirit, M., Wang, C.-C., and Haugh, J. M. 2010. Systematic quantification of negative feedback mechanisms in the extracellular signal-regulated kinase (ERK) signaling network. J Biol Chem, 285(47):36736–44.

Clement, S. L., Scheckel, C., Stoecklin, G., and Lykke-Andersen, J. 2011. Phosphorylation of tristetraprolin by MK2 impairs AU-rich element mRNA decay by preventing deadenylase recruitment. Molecular and Cellular Biology, 31(2): 256–66.

Cole, M. P., Jones, C. T., and Todd, I. D. 1971. A new anti-oestrogenic agent in late breast cancer. An early clinical appraisal of ICI46474. Br J Cancer, 25(2): 270–5.

Consortium, E. P. 2004. The ENCODE (ENCyclopedia Of DNA Elements) Project. Science, 306(5696):636–40.

Cox, A. D., Fesik, S. W., Kimmelman, A. C., Luo, J., and Der, C. J. 2014. Drugging the undruggable RAS: Mission Possible? Nat Rev Drug Discov, 13 (11):828–51.

Crystal, A. S., Shaw, A. T., Sequist, L. V., Friboulet, L., Niederst, M. J., Lockerman, E. L., Frias, R. L., Gainor, J. F., Amzallag, A., Greninger, P., Lee, D., Kalsy, A., Gomez-Caraballo, M., Elamine, L., Howe, E., Hur, W., Lifshits, E., Robinson, H. E., Katayama, R., Faber, A. C., Awad, M. M., Ramaswamy, S., Mino-Kenudson, M., Iafrate, A. J., Benes, C. H., and Engelman, J. A. 2014. Patient-derived models of acquired resistance can identify effective drug combinations for cancer. Science, 346(6216):1480–6.

Dayan, F., Monticelli, M., Pouysségur, J., and Pécou, E. 2009. Gene regulation in response to graded hypoxia: the non-redundant roles of the oxygen sensors PHD and FIH in the HIF pathway. J Theor Biol, 259(2):304–16.

de la Fuente, A., Bing, N., Hoeschele, I., and Mendes, P. 2004. Discovery of meaningful associations in genomic data using partial correlation coefficients. Bioinformatics, 20(18):3565–74.

Demeter, J., Beauheim, C., Gollub, J., Hernandez-Boussard, T., Jin, H., Maier, D., Matese, J. C., Nitzberg, M., Wymore, F., Zachariah, Z. K., Brown, P. O., Sherlock, G., and Ball, C. A. 2007. The Stanford Microarray Database: implementation of new analysis tools and open source release of software. Nucleic Acids Res, 35(Database issue):D766–70.

Destasis. Todesursachen in Deutschland. Fachserie 12 Reihe 4, Statistisches Bundesamt, Wiesbaden, Germany, 2013.

Dhillon, A. S., Hagan, S., Rath, O., and Kolch, W. 2007. MAP kinase signalling pathways in cancer. Oncogene, 26(22):3279–90.

Diaz, L. A., Williams, R. T., Wu, J., Kinde, I., Hecht, J. R., Berlin, J., Allen, B., Bozic, I., Reiter, J. G., Nowak, M. A., Kinzler, K. W., Oliner, K. S., and Vogelstein, B. 2012. The molecular evolution of acquired resistance to targeted EGFR blockade in colorectal cancers. Nature, 486(7404):537–40.

Dougherty, M. K., Müller, J., Ritt, D. A., Zhou, M., Zhou, X. Z., Copeland, T. D., Conrads, T. P., Veenstra, T. D., Lu, K. P., and Morrison, D. K. 2005. Regulation of Raf-1 by direct feedback phosphorylation. Mol Cell, 17(2):215–24.

Douville, E. and Downward, J. 1997. EGF induced SOS phosphorylation in PC12 cells involves P90 RSK-2. Oncogene, 15(4):373–83.

Druker, B. J., Tamura, S., Buchdunger, E., Ohno, S., Segal, G. M., Fanning, S., Zimmermann, J., and Lydon, N. B. 1996. Effects of a selective inhibitor of the Abl tyrosine kinase on the growth of Bcr-Abl positive cells. Nat Med, 2(5):561–6.

Eltzschig, H. K. and Carmeliet, P. 2011. Hypoxia and inflammation. N Engl J Med, 364(7):656–65.

Fähling, M., Persson, A. B., Klinger, B., Benko, E., Steege, A., Kasim, M., Patzak, A., Persson, P. B., Wolf, G., Blüthgen, N., and Mrowka, R. 2012. Multilevel regulation of HIF-1 signaling by TTP. Mol Biol Cell, 23(20):4129–41.

Florini, J. R., Ewton, D. Z., and Coolican, S. A. 1996. Growth hormone and the insulin-like growth factor system in myogenesis. Endocr Rev, 17(5):481–517.

Florkowska, M., Tymoszuk, P., Balwierz, A., Skucha, A., Kochan, J., Wawro, M., Stalinska, K., and Kasza, A. 2012. EGF activates TTP expression by activation of ELK-1 and EGR-1 transcription factors. BMC Mol Biol, 13:8.

Friday, B. B., Yu, C., Dy, G. K., Smith, P. D., Wang, L., Thibodeau, S. N., and Adjei, A. A. 2008. BRAF V600E disrupts AZD6244-induced abrogation of negative feedback pathways between extracellular signal-regulated kinase and Raf proteins. Cancer Res, 68(15):6145–53.

Fritsche-Guenther, R., Witzel, F., Sieber, A., Herr, R., Schmidt, N., Braun, S., Brummer, T., Sers, C., and Blüthgen, N. 2011. Strong negative feedback from Erk to Raf confers robustness to MAPK signalling. Mol Syst Biol, 7:489.

Galbán, S. and Gorospe, M. 2009. Factors interacting with HIF-1alpha mRNA: novel therapeutic targets. Curr Pharm Des, 15(33):3853–60.

Gan, Y., Shi, C., Inge, L., Hibner, M., Balducci, J., and Huang, Y. 2010. Differential roles of ERK and Akt pathways in regulation of EGFR-mediated signaling and motility in prostate cancer cells. Oncogene, 29(35):4947–58.

Gao, J., Aksoy, B. A., Dogrusoz, U., Dresdner, G., Gross, B., Sumer, S. O., Sun, Y., Jacobsen, A., Sinha, R., Larsson, E., Cerami, E., Sander, C., and Schultz, N. 2013. Integrative analysis of complex cancer genomics and clinical profiles using the cBioPortal. Sci Signal, 6(269):pl1.

Gardner, T. S., di Bernardo, D., Lorenz, D., and Collins, J. J. 2003. Inferring genetic networks and identifying compound mode of action via expression profiling. Science, 301(5629):102–5.

Geier, F., Timmer, J., and Fleck, C. 2007. Reconstructing gene-regulatory networks from time series, knock-out data, and prior knowledge. BMC Syst Biol, 1:11.

Gerlinger, M., Rowan, A. J., Horswell, S., Larkin, J., Endesfelder, D., Gronroos, E., Martinez, P., Matthews, N., Stewart, A., Tarpey, P., Varela, I., Phillimore, B., Begum, S., McDonald, N. Q., Butler, A., Jones, D., Raine, K., Latimer, C., Santos, C. R., Nohadani, M., Eklund, A. C., Spencer-Dene, B., Clark, G., Pickering, L., Stamp, G., Gore, M., Szallasi, Z., Downward, J., Futreal, P. A., and Swanton, C. 2012. Intratumor heterogeneity and branched evolution revealed by multiregion sequencing. N Engl J Med, 366(10):883–92.

Goffe, W., Ferrier, G., and Rogers, J. 1994. Global optimization of statistical functions with simulated annealing. Journal of Econometrics, 60(1–2):65–99.

Gopal, Y. N. V., Deng, W., Woodman, S. E., Komurov, K., Ram, P., Smith, P. D., and Davies, M. A. 2010. Basal and treatment-induced activation of AKT mediates resistance to cell death by AZD6244 (ARRY-142886) in Braf-mutant human cutaneous melanoma cells. Cancer Res, 70(21):8736–47.

Griffith, O. L., Montgomery, S. B., Bernier, B., Chu, B., Kasaian, K., Aerts, S., Mahony, S., Sleumer, M. C., Bilenky, M., Haeussler, M., Griffith, M., Gallo, S. M., Giardine, B., Hooghe, B., Van Loo, P., Blanco, E., Ticoll, A., Lithwick, S., Portales-Casamar, E., Donaldson, I. J., Robertson, G., Wadelius, C., De Bleser, P., Vlieghe, D., Halfon, M. S., Wasserman, W., Hardison, R., Bergman, C. M., Jones, S. J. M., and Open Regulatory Annotation Consortium. 2008. ORegAnno:

an open-access community-driven resource for regulatory annotation. Nucleic acids research, 36(Database issue):D107–13.

Guhaniyogi, J. and Brewer, G. 2001. Regulation of mRNA stability in mammalian cells. Gene, 265(1-2):11–23.

Haber, D. and Settleman, J. 2007. Cancer: drivers and passengers. Nature, 446 (7132):145–6.

Halilovic, E., She, Q.-B., Ye, Q., Pagliarini, R., Sellers, W. R., Solit, D. B., and Rosen, N. 2010. PIK3CA mutation uncouples tumor growth and cyclin D1 regulation from MEK/ERK and mutant KRAS signaling. Cancer Res, 70(17): 6804–14.

Hanahan, D. and Coussens, L. M. 2012. Accessories to the crime: functions of cells recruited to the tumor microenvironment. Cancer Cell, 21(3):309–22.

Hanahan, D. and Weinberg, R. A. 2000. The hallmarks of cancer. Cell, 100(1): 57–70.

Hanahan, D. and Weinberg, R. A. 2011. Hallmarks of cancer: the next generation. Cell, 144(5):646–74.

Harris, A. L. 2002. Hypoxia–a key regulatory factor in tumour growth. Nat Rev Cancer, 2(1):38–47.

Harris, S. L. and Levine, A. J. 2005. The p53 pathway: positive and negative feedback loops. Oncogene, 24(17):2899–908.

Hastings, W. 1970. Monte Carlo sampling methods using Markov chains and their applications. Biometrika, 57(1):97–109.

Heiner, M. and Sriram, K. 2010. Structural analysis to determine the core of hypoxia response network. PLoS ONE, 5(1):e8600.

Heinrich, R. and Rapoport, T. A. 1974. A linear steady-state treatment of enzymatic chains. General properties, control and effector strength. Eur J Biochem, 42(1):89–95.

Heiser, L. M., Wang, N. J., Talcott, C. L., Laderoute, K. R., Knapp, M., Guan, Y., Hu, Z., Ziyad, S., Weber, B. L., Laquerre, S., Jackson, J. R., Wooster, R. F., Kuo, W. L., Gray, J. W., and Spellman, P. T. 2009. Integrated analysis of breast cancer cell lines reveals unique signaling pathways. Genome Biol, 10(3):R31.

Herbst, R. S., Fukuoka, M., and Baselga, J. 2004. Gefitinib–a novel targeted approach to treating cancer. Nat Rev Cancer, 4(12):956–65.

Hitti, E., Iakovleva, T., Brook, M., Deppenmeier, S., Gruber, A. D., Radzioch, D., Clark, A. R., Blackshear, P. J., Kotlyarov, A., and Gaestel, M. 2006. Mitogen-activated protein kinase-activated protein kinase 2 regulates tumor necrosis factor mRNA stability and translation mainly by altering tristetraprolin expression, stability, and binding to adenine/uridine-rich element. Mol Cell Biol, 26(6):2399–407.

Holohan, C., Schaeybroeck, S. V., Longley, D. B., and Johnston, P. G. 2013. Cancer drug resistance: an evolving paradigm. Nat Rev Cancer, 13(10):714–26.

Hudziak, R. M., Lewis, G. D., Winget, M., Fendly, B. M., Shepard, H. M., and Ullrich, A. 1989. p185HER2 monoclonal antibody has antiproliferative effects in vitro and sensitizes human breast tumor cells to tumor necrosis factor. Mol Cell Biol, 9(3):1165–72.

Huggins, C. and Stevens, R. 1941. Studies on prostatic cancer: II. The effects of castration on advanced carcinoma of the prostate gland. Archives of surgery.

Ideker, T., Galitski, T., and Hood, L. 2001. A new approach to decoding life: systems biology. Annu Rev Genomics Hum Genet, 2:343–72.

Iwasa, Y., Nowak, M. A., and Michor, F. 2006. Evolution of resistance during clonal expansion. Genetics, 172(4):2557–66.

Jemal, A., Ward, E., and Thun, M. 2010. Declining death rates reflect progress against cancer. PLoS ONE, 5(3):e9584.

Jones, D. S., Podolsky, S. H., and Greene, J. A. 2012. The burden of disease and the changing task of medicine. N Engl J Med, 366(25):2333–8.

Kacser, H. and Burns, J. A. 1973. The control of flux. Symp Soc Exp Biol, 27: 65–104.

Kaelin, W. G. 2005. The concept of synthetic lethality in the context of anticancer therapy. Nat Rev Cancer, 5(9):689–98.

Kandoth, C., McLellan, M. D., Vandin, F., Ye, K., Niu, B., Lu, C., Xie, M., Zhang, Q., McMichael, J. F., Wyczalkowski, M. A., Leiserson, M. D. M., Miller, C. A., Welch, J. S., Walter, M. J., Wendl, M. C., Ley, T. J., Wilson, R. K., Raphael,

B. J., and Ding, L. 2013. Mutational landscape and significance across 12 major cancer types. Nature, 502(7471):333–9.

Kanehisa, M. and Goto, S. 2000. KEGG: kyoto encyclopedia of genes and genomes. Nucleic Acids Res, 28(1):27–30.

Karapetis, C. S., Khambata-Ford, S., Jonker, D. J., O'Callaghan, C. J., Tu, D., Tebbutt, N. C., Simes, R. J., Chalchal, H., Shapiro, J. D., Robitaille, S., Price, T. J., Shepherd, L., Au, H.-J., Langer, C., Moore, M. J., and Zalcberg, J. R. 2008. K-ras mutations and benefit from cetuximab in advanced colorectal cancer. N Engl J Med, 359(17):1757–65.

Karin, M., Yamamoto, Y., and Wang, Q. M. 2004. The IKK NF-kappa B system: a treasure trove for drug development. Nat Rev Drug Discov, 3(1):17–26.

Kholodenko, B., Yaffe, M. B., and Kolch, W. 2012. Computational approaches for analyzing information flow in biological networks. Sci Signal, 5(220):re1.

Kholodenko, B. N. 2000. Negative feedback and ultrasensitivity can bring about oscillations in the mitogen-activated protein kinase cascades. Eur J Biochem, 267(6):1583–8.

Kholodenko, B. N., Hoek, J. B., Westerhoff, H. V., and Brown, G. C. 1997. Quantification of information transfer via cellular signal transduction pathways. FEBS Lett, 414(2):430–4.

Kholodenko, B. N., Demin, O. V., Moehren, G., and Hoek, J. B. 1999. Quantification of short term signaling by the epidermal growth factor receptor. J Biol Chem, 274(42):30169–81.

Kholodenko, B. N., Kiyatkin, A., Bruggeman, F. J., Sontag, E., Westerhoff, H. V., and Hoek, J. B. 2002. Untangling the wires: a strategy to trace functional interactions in signaling and gene networks. Proc Natl Acad Sci USA, 99(20): 12841–6.

Kholodenko, B. N., Hancock, J. F., and Kolch, W. 2010. Signalling ballet in space and time. Nat Rev Mol Cell Biol, 11(6):414–26.

Khvalevsky, E. Z., Gabai, R., Rachmut, I. H., Horwitz, E., Brunschwig, Z., Orbach, A., Shemi, A., Golan, T., Domb, A. J., Yavin, E., Giladi, H., Rivkin, L., Simerzin, A., Eliakim, R., Khalaileh, A., Hubert, A., Lahav, M., Kopelman, Y., Goldin, E., Dancour, A., Hants, Y., Arbel-Alon, S., Abramovitch, R., Shemi, A., and

Galun, E. 2013. Mutant KRAS is a druggable target for pancreatic cancer. Proc Natl Acad Sci USA, 110(51):20723–28.

Kim, D., Rath, O., Kolch, W., and Cho, K.-H. 2007. A hidden oncogenic positive feedback loop caused by crosstalk between Wnt and ERK pathways. Oncogene, 26(31):4571–9.

Kim, J.-W. and Dang, C. V. 2006. Cancer's molecular sweet tooth and the Warburg effect. Cancer Res, 66(18):8927–30.

Kim, T. W., Yim, S., Choi, B. J., Jang, Y., Lee, J. J., Sohn, B. H., Yoo, H.-S., Yeom, Y. I., and Park, K. C. 2010. Tristetraprolin regulates the stability of HIF-1alpha mRNA during prolonged hypoxia. Biochemical and Biophysical Research Communications, 391(1):963–8.

Kitano, H. 2002. Systems biology: a brief overview. Science, 295(5560):1662–4.

Klinger, B. and Blüthgen, N. 2014. Consequences of feedback in signal transduction for targeted therapies. Biochem Soc Trans, 42(4):770–5.

Klinger, B., Sieber, A., Fritsche-Guenther, R., Witzel, F., Berry, L., Schumacher, D., Yan, Y., Durek, P., Merchant, M., Schäfer, R., Sers, C., and Blüthgen, N. 2013. Network quantification of EGFR signaling unveils potential for targeted combination therapy. Mol Syst Biol, 9:673.

Kohn, K. W., Riss, J., Aprelikova, O., Weinstein, J. N., Pommier, Y., and Barrett, J. C. 2004. Properties of switch-like bioregulatory networks studied by simulation of the hypoxia response control system. Mol Biol Cell, 15(7):3042–52.

Kolch, W., Calder, M., and Gilbert, D. 2005. When kinases meet mathematics: the systems biology of MAPK signalling. FEBS Lett, 579(8):1891–5.

König, J., Zarnack, K., Luscombe, N. M., and Ule, J. 2011. Protein-RNA interactions: new genomic technologies and perspectives. Nat Rev Genet, 13(2): 77–83.

Koppenol, W. H., Bounds, P. L., and Dang, C. V. 2011. Otto Warburg's contributions to current concepts of cancer metabolism. Nat Rev Cancer, 11(5): 325–37.

Kreeger, P. K. and Lauffenburger, D. A. 2010. Cancer systems biology: a network modeling perspective. Carcinogenesis, 31(1):2–8.

Kreeger, P. K., Mandhana, R., Alford, S. K., Haigis, K. M., and Lauffenburger, D. A. 2009. RAS mutations affect tumor necrosis factor-induced apoptosis in colon carcinoma cells via ERK-modulatory negative and positive feedback circuits along with non-ERK pathway effects. Cancer Res, 69(20):8191–9.

Küffner, R., Petri, T., Windhager, L., and Zimmer, R. 2010. Petri Nets with Fuzzy Logic (PNFL): reverse engineering and parametrization. PLoS ONE, 5(9).

Kuschel, A., Simon, P., and Tug, S. 2012. Functional regulation of HIF-1α under normoxia–is there more than post-translational regulation? J Cell Physiol, 227 (2):514–24.

Lai, A. Z., Cory, S., Zhao, H., Gigoux, M., Monast, A., Guiot, M.-C., Huang, S., Tofigh, A., Thompson, C., Naujokas, M., Marcus, V. A., Bertos, N., Sehat, B., Perera, R. M., Bell, E. S., Page, B. D. G., Gunning, P. T., Ferri, L. E., Hallett, M., and Park, M. 2014. Dynamic reprogramming of signaling upon met inhibition reveals a mechanism of drug resistance in gastric cancer. Sci Signal, 7 (322):ra38.

Lau, K. S., Juchheim, A. M., Cavaliere, K. R., Philips, S. R., Lauffenburger, D. A., and Haigis, K. M. 2011. In vivo systems analysis identifies spatial and temporal aspects of the modulation of TNF-α-induced apoptosis and proliferation by MAPKs. Sci Signal, 4(165):ra16.

Lee, M. J., Ye, A. S., Gardino, A. K., Heijink, A. M., Sorger, P. K., MacBeath, G., and Yaffe, M. B. 2012. Sequential application of anticancer drugs enhances cell death by rewiring apoptotic signaling networks. Cell, 149(4):780–94.

Lee-Fruman, K. K., Kuo, C. J., Lippincott, J., Terada, N., and Blenis, J. 1999. Characterization of S6K2, a novel kinase homologous to S6K1. Oncogene, 18 (36):5108–14.

Legewie, S., Herzel, H., Westerhoff, H. V., and Blüthgen, N. 2008. Recurrent design patterns in the feedback regulation of the mammalian signalling network. Mol Syst Biol, 4:190.

Lehman, J. A. and Gomez-Cambronero, J. 2002. Molecular crosstalk between p70S6k and MAPK cell signaling pathways. Biochem Biophys Res Commun, 293(1):463–9.

Lourakis, M. levmar: Levenberg-Marquardt nonlinear least squares algorithms in C/C++. [web page] `http://www.ics.forth.gr/~lourakis/levmar/`, Jul. 2004. [Accessed on 31 Jan. 2005.].

Lu, Y., Muller, M., Smith, D., Dutta, B., Komurov, K., Iadevaia, S., Ruths, D., Tseng, J. T., Yu, S., Yu, Q., Nakhleh, L., Balazsi, G., Donnelly, J., Schurdak, M., Morgan-Lappe, S., Fesik, S., Ram, P. T., and Mills, G. B. 2011. Kinome siRNA-phosphoproteomic screen identifies networks regulating AKT signaling. Oncogene, 30(45):4567–77.

Luo, J., Solimini, N. L., and Elledge, S. J. 2009. Principles of cancer therapy: oncogene and non-oncogene addiction. Cell, 136(5):823–37.

Mader, R. M., Foerster, S., Sarin, N., Michaelis, M., Cinatl, J., Kloft, C., Fröhlich, H., Engel, F., Kalayda, G. V., Jäger, W., Frötschl, R., Jaehde, U., and Ritter, C. A. 2014. NSCLC cells adapted to EGFR inhibition accumulate EGFR interacting proteins and down-regulate microRNA related to epithelial-mesenchymal transition. Int J Clin Pharmacol Ther, 52(1):92–4.

Marbach, D., Schaffter, T., Mattiussi, C., and Floreano, D. 2009. Generating realistic in silico gene networks for performance assessment of reverse engineering methods. Journal of Computational Biology, 16(2):229–39.

Marshall, C. J. 1995. Specificity of receptor tyrosine kinase signaling: transient versus sustained extracellular signal-regulated kinase activation. Cell, 80(2):179–85.

Mason, J. M., Morrison, D. J., Basson, M. A., and Licht, J. D. 2006. Sprouty proteins: multifaceted negative-feedback regulators of receptor tyrosine kinase signaling. Trends Cell Biol, 16(1):45–54.

Massagué, J. 2012. TGFβ signalling in context. Nat Rev Mol Cell Biol, 13(10): 616–30.

Matys, V., Kel-Margoulis, O. V., Fricke, E., Liebich, I., Land, S., Barre-Dirrie, A., Reuter, I., Chekmenev, D., Krull, M., Hornischer, K., Voss, N., Stegmaier, P., Lewicki-Potapov, B., Saxel, H., Kel, A. E., and Wingender, E. 2006. TRANSFAC and its module TRANSCompel: transcriptional gene regulation in eukaryotes. Nucleic acids research, 34(Database issue):D108–10.

Mazzarello, P. 1999. A unifying concept: the history of cell theory. Nat Cell Biol, 1(1):E13–5.

Mbalaviele, G., Sommers, C. D., Bonar, S. L., Mathialagan, S., Schindler, J. F., Guzova, J. A., Shaffer, A. F., Melton, M. A., Christine, L. J., Tripp, C. S., Chiang, P.-C., Thompson, D. C., Hu, Y., and Kishore, N. 2009. A novel, highly selective, tight binding IkappaB kinase-2 (IKK-2) inhibitor: a tool to correlate IKK-2 activity to the fate and functions of the components of the nuclear factor-kappaB pathway in arthritis-relevant cells and animal models. J Pharmacol Exp Ther, 329(1):14–25.

McKay, M. D., Beckman, R. J., and Conover, W. J. 1979. A comparison of three methods for selecting values of input variables in the analysis of output from a computer code. Technometrics, 21(2):239–245.

Mendes, P., Sha, W., and Ye, K. 2003. Artificial gene networks for objective comparison of analysis algorithms. Bioinformatics, 19 Suppl 2:ii122–9.

Menzies, A. M. and Long, G. V. 2014. Dabrafenib and trametinib, alone and in combination for BRAF-mutant metastatic melanoma. Clin Cancer Res, 20(8): 2035–43.

Miller, M. L., Molinelli, E. J., Nair, J. S., Sheikh, T., Samy, R., Jing, X., He, Q., Korkut, A., Crago, A. M., Singer, S., Schwartz, G. K., and Sander, C. 2013. Drug synergy screen and network modeling in dedifferentiated liposarcoma identifies CDK4 and IGF1R as synergistic drug targets. Sci Signal, 6(294):ra85.

Mirzoeva, O. K., Das, D., Heiser, L. M., Bhattacharya, S., Siwak, D., Gendelman, R., Bayani, N., Wang, N. J., Neve, R. M., Guan, Y., Hu, Z., Knight, Z., Feiler, H. S., Gascard, P., Parvin, B., Spellman, P. T., Shokat, K. M., Wyrobek, A. J., Bissell, M. J., McCormick, F., Kuo, W.-L., Mills, G. B., Gray, J. W., and Korn, W. M. 2009. Basal subtype and MAPK/ERK kinase (MEK)-phosphoinositide 3-kinase feedback signaling determine susceptibility of breast cancer cells to MEK inhibition. Cancer Res, 69(2):565–72.

Misale, S., Yaeger, R., Hobor, S., Scala, E., Janakiraman, M., Liska, D., Valtorta, E., Schiavo, R., Buscarino, M., Siravegna, G., Bencardino, K., Cercek, A., Chen, C.-T., Veronese, S., Zanon, C., Sartore-Bianchi, A., Gambacorta, M., Gallicchio, M., Vakiani, E., Boscaro, V., Medico, E., Weiser, M., Siena, S., Nicolantonio, F. D., Solit, D., and Bardelli, A. 2012. Emergence of KRAS mutations and acquired resistance to anti-EGFR therapy in colorectal cancer. Nature, 486 (7404):532–6.

Misale, S., Arena, S., Lamba, S., Siravegna, G., Lallo, A., Hobor, S., Russo, M., Buscarino, M., Lazzari, L., Sartore-Bianchi, A., Bencardino, K., Amatu, A., Lauricella, C., Valtorta, E., Siena, S., Nicolantonio, F. D., and Bardelli, A. 2014. Blockade of EGFR and MEK Intercepts Heterogeneous Mechanisms of

Acquired Resistance to Anti-EGFR Therapies in Colorectal Cancer. Sci Transl Med, 6(224):224ra26.

Molinelli, E. J., Korkut, A., Wang, W., Miller, M. L., Gauthier, N. P., Jing, X., Kaushik, P., He, Q., Mills, G., Solit, D. B., Pratilas, C. A., Weigt, M., Braunstein, A., Pagnani, A., Zecchina, R., and Sander, C. 2013. Perturbation biology: inferring signaling networks in cellular systems. PLoS Comput Biol, 9 (12):e1003290.

Morris, M. K., Saez-Rodriguez, J., Clarke, D. C., Sorger, P. K., and Lauffenburger, D. A. 2011. Training signaling pathway maps to biochemical data with constrained fuzzy logic: quantitative analysis of liver cell responses to inflammatory stimuli. PLoS Comput Biol, 7(3):e1001099.

Moyer, J. D., Barbacci, E. G., Iwata, K. K., Arnold, L., Boman, B., Cunningham, A., DiOrio, C., Doty, J., Morin, M. J., Moyer, M. P., Neveu, M., Pollack, V. A., Pustilnik, L. R., Reynolds, M. M., Sloan, D., Theleman, A., and Miller, P. 1997. Induction of apoptosis and cell cycle arrest by CP-358,774, an inhibitor of epidermal growth factor receptor tyrosine kinase. Cancer Res, 57(21):4838–48.

Mukherjee, S. The Emperor of All Maladies. Fourth Estate, London, 2011. ISBN 978-0-00-725092-9.

Müller, B. and Grossniklaus, U. 2010. Model organisms–A historical perspective. J Proteomics, 73(11):2054–63.

Murphy, L. O., Smith, S., Chen, R.-H., Fingar, D. C., and Blenis, J. 2002. Molecular interpretation of ERK signal duration by immediate early gene products. Nature cell biology, 4(8):556–564.

Murphy, L. O., MacKeigan, J. P., and Blenis, J. 2004. A network of immediate early gene products propagates subtle differences in mitogen-activated protein kinase signal amplitude and duration. Molecular and cellular biology, 24(1): 144–153.

Nelander, S., Wang, W., Nilsson, B., She, Q.-B., Pratilas, C., Rosen, N., Gennemark, P., and Sander, C. 2008. Models from experiments: combinatorial drug perturbations of cancer cells. Molecular Systems Biology, 4:216.

Novère, N. L. 2015. Quantitative and logic modelling of molecular and gene networks. Nat Rev Genet, 16:146–58.

Oda, K., Matsuoka, Y., Funahashi, A., and Kitano, H. 2005. A comprehensive pathway map of epidermal growth factor receptor signaling. Mol Syst Biol, 1: 2005.0010.

Odom, D. T., Zizlsperger, N., Gordon, D. B., Bell, G. W., Rinaldi, N. J., Murray, H. L., Volkert, T. L., Schreiber, J., Rolfe, P. A., Gifford, D. K., Fraenkel, E., Bell, G. I., and Young, R. A. 2004. Control of pancreas and liver gene expression by HNF transcription factors. Science, 303(5662):1378–81.

Oeckinghaus, A., Hayden, M. S., and Ghosh, S. 2011. Crosstalk in NF-kappa B signaling pathways. Nat Immunol, 12(8):695–708.

Ogawa, K., Chen, F., Kim, Y.-J., and Chen, Y. 2003. Transcriptional regulation of tristetraprolin by transforming growth factor-beta in human T cells. J Biol Chem, 278(32):30373–81.

Ostrem, J. M., Peters, U., Sos, M. L., Wells, J. A., and Shokat, K. M. 2013. K-Ras(G12C) inhibitors allosterically control GTP affinity and effector interactions. Nature, 503(7477):548–51.

Owen, A. B. 1992. A central limit theorem for Latin hypercube sampling. J. Roy. Statist. Soc. Ser. B, 54(2):541–551.

Parikh, C., Janakiraman, V., Wu, W.-I., Foo, C. K., Kljavin, N. M., Chaudhuri, S., Stawiski, E., Lee, B., Lin, J., Li, H., Lorenzo, M. N., Yuan, W., Guillory, J., Jackson, M., Rondon, J., Franke, Y., Bowman, K. K., Sagolla, M., Stinson, J., Wu, T. D., Wu, J., Stokoe, D., Stern, H. M., Brandhuber, B. J., Lin, K., Skelton, N. J., and Seshagiri, S. 2012. Disruption of PH-kinase domain interactions leads to oncogenic activation of AKT in human cancers. Proc Natl Acad Sci USA, 109 (47):19368–73.

Parikh, J. R., Klinger, B., Xia, Y., Marto, J. A., and Blüthgen, N. 2010. Discovering causal signaling pathways through gene-expression patterns. Nucleic Acids Res, 38 Suppl:W109–17.

Piccart, M. 2008. Circumventing de novo and acquired resistance to trastuzumab: new hope for the care of ErbB2-positive breast cancer. Clin Breast Cancer, 8 Suppl 3:S100–13.

Plongthongkum, N., Diep, D. H., and Zhang, K. 2014. Advances in the profiling of DNA modifications: cytosine methylation and beyond. Nat Rev Genet, 15(10): 647–61.

Popper, K. 1957. Philosophy of science: A personal report. British philosophy in the mid-century.

Porter, A. C. and Vaillancourt, R. R. 1998. Tyrosine kinase receptor-activated signal transduction pathways which lead to oncogenesis. Oncogene, 17(11 Reviews): 1343–52.

Poulikakos, P. I. and Solit, D. B. 2011. Resistance to MEK inhibitors: should we co-target upstream? Sci Signal, 4(166):pe16.

Poulikakos, P. I., Persaud, Y., Janakiraman, M., Kong, X., Ng, C., Moriceau, G., Shi, H., Atefi, M., Titz, B., Gabay, M. T., Salton, M., Dahlman, K. B., Tadi, M., Wargo, J. A., Flaherty, K. T., Kelley, M. C., Misteli, T., Chapman, P. B., Sosman, J. A., Graeber, T. G., Ribas, A., Lo, R. S., Rosen, N., and Solit, D. B. 2011. RAF inhibitor resistance is mediated by dimerization of aberrantly spliced BRAF(V600E). Nature, 480(7377):387–90.

Pouysségur, J., Dayan, F., and Mazure, N. M. 2006. Hypoxia signalling in cancer and approaches to enforce tumour regression. Nature, 441(7092):437–43.

Prahallad, A., Sun, C., Huang, S., Nicolantonio, F. D., Salazar, R., Zecchin, D., Beijersbergen, R. L., Bardelli, A., and Bernards, R. 2012. Unresponsiveness of colon cancer to BRAF(V600E) inhibition through feedback activation of EGFR. Nature, 483(7387):100–3.

Pratilas, C. A., Taylor, B. S., Ye, Q., Viale, A., Sander, C., Solit, D. B., and Rosen, N. 2009. (V600E)BRAF is associated with disabled feedback inhibition of RAF-MEK signaling and elevated transcriptional output of the pathway. Proc Natl Acad Sci USA, 106(11):4519–24.

Prenen, H., Tejpar, S., and Cutsem, E. V. 2010. New strategies for treatment of KRAS mutant metastatic colorectal cancer. Clin Cancer Res, 16(11):2921–6.

Qutub, A. A. and Popel, A. S. 2006. A computational model of intracellular oxygen sensing by hypoxia-inducible factor HIF1 alpha. J Cell Sci, 119(Pt 16):3467–80.

Rad, R., Cadiñanos, J., Rad, L., Varela, I., Strong, A., Kriegl, L., Constantino-Casas, F., Eser, S., Hieber, M., Seidler, B., Price, S., Fraga, M. F., Calvanese, V., Hoffman, G., Ponstingl, H., Schneider, G., Yusa, K., Grove, C., Schmid, R. M., Wang, W., Vassiliou, G., Kirchner, T., McDermott, U., Liu, P., Saur, D., and Bradley, A. 2013. A genetic progression model of Braf(V600E)-induced intestinal tumorigenesis reveals targets for therapeutic intervention. Cancer Cell, 24(1):15–29.

Raue, A., Kreutz, C., Maiwald, T., Bachmann, J., Schilling, M., Klingmüller, U., and Timmer, J. 2009. Structural and practical identifiability analysis of partially observed dynamical models by exploiting the profile likelihood. Bioinformatics, 25(15):1923–9.

Roberts, P. J. and Der, C. J. 2007. Targeting the Raf-MEK-ERK mitogen-activated protein kinase cascade for the treatment of cancer. Oncogene, 26(22):3291–310.

Ross, C. R., Brennan-Laun, S. E., and Wilson, G. M. 2012. Tristetraprolin: roles in cancer and senescence. Ageing Res Rev, 11(4):473–84.

Rossignol, F., Vaché, C., and Clottes, E. 2002. Natural antisense transcripts of hypoxia-inducible factor 1alpha are detected in different normal and tumour human tissues. Gene, 299(1-2):135–40.

Roth, A. D., Tejpar, S., Delorenzi, M., Yan, P., Fiocca, R., Klingbiel, D., Dietrich, D., Biesmans, B., Bodoky, G., Barone, C., Aranda, E., Nordlinger, B., Cisar, L., Labianca, R., Cunningham, D., Cutsem, E. V., and Bosman, F. 2010. Prognostic role of KRAS and BRAF in stage II and III resected colon cancer: results of the translational study on the PETACC-3, EORTC 40993, SAKK 60-00 trial. J Clin Oncol, 28(3):466–74.

Saez-Rodriguez, J., Alexopoulos, L. G., Epperlein, J., Samaga, R., Lauffenburger, D. A., Klamt, S., and Sorger, P. K. 2009. Discrete logic modelling as a means to link protein signalling networks with functional analysis of mammalian signal transduction. Molecular Systems Biology, 5:331.

Saez-Rodriguez, J., Alexopoulos, L. G., Zhang, M., Morris, M. K., Lauffenburger, D. A., and Sorger, P. K. 2011. Comparing signaling networks between normal and transformed hepatocytes using discrete logical models. Cancer Res, 71(16): 5400–11.

Sahin, O., Fröhlich, H., Löbke, C., Korf, U., Burmester, S., Majety, M., Mattern, J., Schupp, I., Chaouiya, C., Thieffry, D., Poustka, A., Wiemann, S., Beissbarth, T., and Arlt, D. 2009. Modeling ERBB receptor-regulated G1/S transition to find novel targets for de novo trastuzumab resistance. BMC Syst Biol, 3:1.

Sandler, H. and Stoecklin, G. 2008. Control of mRNA decay by phosphorylation of tristetraprolin. Biochem. Soc. Trans, 36(Pt 3):491–6.

Santos, S. D. M., Verveer, P. J., and Bastiaens, P. I. H. 2007. Growth factor-induced MAPK network topology shapes Erk response determining PC-12 cell fate. Nat Cell Biol, 9(3):324–30.

Santra, T., Kolch, W., and Kholodenko, B. N. 2013. Integrating Bayesian variable selection with Modular Response Analysis to infer biochemical network topology. BMC Syst Biol, 7:57.

Schelker, M., Raue, A., Timmer, J., and Kreutz, C. 2012. Comprehensive estimation of input signals and dynamics in biochemical reaction networks. Bioinformatics, 28(18):i529–34.

Schmierer, B., Novák, B., and Schofield, C. J. 2010. Hypoxia-dependent sequestration of an oxygen sensor by a widespread structural motif can shape the hypoxic response–a predictive kinetic model. BMC Syst Biol, 4:139.

Schoeberl, B., Eichler-Jonsson, C., Gilles, E. D., and Müller, G. 2002. Computational modeling of the dynamics of the MAP kinase cascade activated by surface and internalized EGF receptors. Nat Biotechnol, 20(4):370–5.

Schulthess, P. and Blüthgen, N. 2011. From reaction networks to information flow–using modular response analysis to track information in signaling networks. Meth Enzymol, 500:397–409.

Schwanhäusser, B., Busse, D., Li, N., Dittmar, G., Schuchhardt, J., Wolf, J., Chen, W., and Selbach, M. 2011. Global quantification of mammalian gene expression control. Nature, 473(7347):337–42.

Sebens, S. and Schafer, H. 2012. The tumor stroma as mediator of drug resistance–a potential target to improve cancer therapy? Curr Pharm Biotechnol, 13(11): 2259–72.

Semenza, G. L. 2003. Targeting HIF-1 for cancer therapy. Nat Rev Cancer, 3(10): 721–32.

Semenza, G. L. 2010a. HIF-1: upstream and downstream of cancer metabolism. Curr Opin Genet Dev, 20(1):51–6.

Semenza, G. L. 2010b. Defining the role of hypoxia-inducible factor 1 in cancer biology and therapeutics. Oncogene, 29(5):625–34.

Serra, V., Scaltriti, M., Prudkin, L., Eichhorn, P. J. A., Ibrahim, Y. H., Chandarlapaty, S., Markman, B., Rodriguez, O., Guzman, M., Rodriguez, S., Gili,

M., Russillo, M., Parra, J. L., Singh, S., Arribas, J., Rosen, N., and Baselga, J. 2011. PI3K inhibition results in enhanced HER signaling and acquired ERK dependency in HER2-overexpressing breast cancer. Oncogene, 30(22):2547–57.

Shankaran, H. and Wiley, H. S. 2010. Oscillatory dynamics of the extracellular signal-regulated kinase pathway. Curr Opin Genet Dev, 20(6):650–5.

Shankavaram, U. T., Reinhold, W. C., Nishizuka, S., Major, S., Morita, D., Chary, K. K., Reimers, M. A., Scherf, U., Kahn, A., Dolginow, D., Cossman, J., Kaldjian, E. P., Scudiero, D. A., Petricoin, E., Liotta, L., Lee, J. K., and Weinstein, J. N. 2007. Transcript and protein expression profiles of the NCI-60 cancer cell panel: an integromic microarray study. Mol Cancer Ther, 6(3):820–32.

Sharova, L. V., Sharov, A. A., Nedorezov, T., Piao, Y., Shaik, N., and Ko, M. S. H. 2009. Database for mRNA half-life of 19 977 genes obtained by DNA microarray analysis of pluripotent and differentiating mouse embryonic stem cells. DNA Res, 16(1):45–58.

She, Q.-B., Solit, D. B., Ye, Q., O'Reilly, K. E., Lobo, J., and Rosen, N. 2005. The BAD protein integrates survival signaling by EGFR/MAPK and PI3K/Akt kinase pathways in PTEN-deficient tumor cells. Cancer Cell, 8(4):287–97.

Shtivelman, E., Lifshitz, B., Gale, R. P., and Canaani, E. 1985. Fused transcript of abl and bcr genes in chronic myelogenous leukaemia. Nature, 315(6020):550–4.

Siegel, R., Naishadham, D., and Jemal, A. 2013. Cancer statistics, 2013. CA Cancer J Clin, 63(1):11–30.

Singer, C. 1914. Notes on the Early History of Microscopy. Proc R Soc Med, 7 (Sect Hist Med):247–79.

Solit, D. B., Garraway, L. A., Pratilas, C. A., Sawai, A., Getz, G., Basso, A., Ye, Q., Lobo, J. M., She, Y., Osman, I., Golub, T. R., Sebolt-Leopold, J., Sellers, W. R., and Rosen, N. 2006. BRAF mutation predicts sensitivity to MEK inhibition. Nature, 439(7074):358–62.

Stark, J., Brewer, D., Barenco, M., Tomescu, D., Callard, R., and Hubank, M. 2003. Reconstructing gene networks: what are the limits? Biochemical Society Transactions, 31(Pt 6):1519–25.

Staudacher, J. J., de Vries, I. S. N., Ujvari, S. J., Klinger, B., Kasim, M., Benko, E., Ostareck-Lederer, A., Ostareck, D. H., Persson, A. B., Lorenzen, S., Meier, J. C.,

Blüthgen, N., Persson, P. B., Henrion-Caude, A., Mrowka, R., and Fähling, M. 2015. Hypoxia-induced gene expression results from selective mRNA partitioning to the endoplasmic reticulum. Nucleic Acids Res, 43(6):3219–36.

Stelniec-Klotz, I., Legewie, S., Tchernitsa, O., Witzel, F., Klinger, B., Sers, C., Herzel, H., Blüthgen, N., and Schäfer, R. 2012. Reverse engineering a hierarchical regulatory network downstream of oncogenic KRAS. Molecular Systems Biology, 8:601.

Stoecklin, G. and Anderson, P. 2007. In a tight spot: ARE-mRNAs at processing bodies. Genes Dev, 21(6):627–31.

Stoecklin, G., Stubbs, T., Kedersha, N., Wax, S., Rigby, W. F. C., Blackwell, T. K., and Anderson, P. 2004. MK2-induced tristetraprolin:14-3-3 complexes prevent stress granule association and ARE-mRNA decay. EMBO J, 23(6):1313–24.

Stolovitzky, G., Monroe, D., and Califano, A. 2007. Dialogue on reverse-engineering assessment and methods: the DREAM of high-throughput pathway inference. Ann N Y Acad Sci, 1115:1–22.

Stolovitzky, G., Prill, R. J., and Califano, A. 2009. Lessons from the DREAM2 Challenges. Ann N Y Acad Sci, 1158:159–95.

Straussman, R., Morikawa, T., Shee, K., Barzily-Rokni, M., Qian, Z. R., Du, J., Davis, A., Mongare, M. M., Gould, J., Frederick, D. T., Cooper, Z. A., Chapman, P. B., Solit, D. B., Ribas, A., Lo, R. S., Flaherty, K. T., Ogino, S., Wargo, J. A., and Golub, T. R. 2012. Tumour micro-environment elicits innate resistance to RAF inhibitors through HGF secretion. Nature, 487(7408):500–4.

Sturm, O. E., Orton, R., Grindlay, J., Birtwistle, M., Vyshemirsky, V., Gilbert, D., Calder, M., Pitt, A., Kholodenko, B., and Kolch, W. 2010. The mammalian MAPK/ERK pathway exhibits properties of a negative feedback amplifier. Sci Signal, 3(153):ra90.

Subramanian, A., Tamayo, P., Mootha, V. K., Mukherjee, S., Ebert, B. L., Gillette, M. A., Paulovich, A., Pomeroy, S. L., Golub, T. R., Lander, E. S., and Mesirov, J. P. 2005. Gene set enrichment analysis: a knowledge-based approach for interpreting genome-wide expression profiles. Proc Natl Acad Sci USA, 102(43): 15545–50.

Sun, L., Stoecklin, G., Way, S. V., Hinkovska-Galcheva, V., Guo, R.-F., Anderson, P., and Shanley, T. P. 2007. Tristetraprolin (TTP)-14-3-3 complex formation protects TTP from dephosphorylation by protein phosphatase 2a and stabilizes tumor necrosis factor-alpha mRNA. J Biol Chem, 282(6):3766–77.

Tchen, C. R., Brook, M., Saklatvala, J., and Clark, A. R. 2004. The stability of tristetraprolin mRNA is regulated by mitogen-activated protein kinase p38 and by tristetraprolin itself. J Biol Chem, 279(31):32393–400.

Tchernitsa, O. I., Sers, C., Zuber, J., Hinzmann, B., Grips, M., Schramme, A., Lund, P., Schwendel, A., Rosenthal, A., and Schäfer, R. 2004. Transcriptional basis of KRAS oncogene-mediated cellular transformation in ovarian epithelial cells. Oncogene, 23(26):4536–55.

Tentner, A. R., Lee, M. J., Ostheimer, G. J., Samson, L. D., Lauffenburger, D. A., and Yaffe, M. B. 2012. Combined experimental and computational analysis of DNA damage signaling reveals context-dependent roles for Erk in apoptosis and G1/S arrest after genotoxic stress. Mol Syst Biol, 8:568.

Thomas, P., Durek, P., Solt, I., Klinger, B., Witzel, F., Schulthess, P., Mayer, Y., Tikk, D., Blüthgen, N., and Leser, U. 2015. Computer-assisted curation of a human regulatory core network from the biological literature. Bioinformatics, 31 (8):1258–66.

Tian, Q., Stepaniants, S. B., Mao, M., Weng, L., Feetham, M. C., Doyle, M. J., Yi, E. C., Dai, H., Thorsson, V., Eng, J., Goodlett, D., Berger, J. P., Gunter, B., Linseley, P. S., Stoughton, R. B., Aebersold, R., Collins, S. J., Hanlon, W. A., and Hood, L. E. 2004. Integrated genomic and proteomic analyses of gene expression in Mammalian cells. Mol Cell Proteomics, 3(10):960–9.

Timmer, J., Rust, H., Horbelt, W., and Voss, H. 2000. Parametric, nonparametric and parametric modelling of a chaotic circuit time series. Physics Letters A, 274 (3-4):123–34.

Timmer, J., Müller, T., Swameye, I., Sandra, O., and Klingm"uller, U. 2004. Modeling the nonlinear dynamics of cellular signal transduction. Int. J. Bifurcation Chaos, 14(06):2069–2079.

Trédan, O., Galmarini, C. M., Patel, K., and Tannock, I. F. 2007. Drug resistance and the solid tumor microenvironment. J Natl Cancer Inst, 99(19):1441–54.

Tsai, J., Lee, J. T., Wang, W., Zhang, J., Cho, H., Mamo, S., Bremer, R., Gillette, S., Kong, J., Haass, N. K., Sproesser, K., Li, L., Smalley, K. S. M., Fong, D., Zhu, Y.-L., Marimuthu, A., Nguyen, H., Lam, B., Liu, J., Cheung, I., Rice, J., Suzuki,

Y., Luu, C., Settachatgul, C., Shellooe, R., Cantwell, J., Kim, S.-H., Schlessinger, J., Zhang, K. Y. J., West, B. L., Powell, B., Habets, G., Zhang, C., Ibrahim, P. N., Hirth, P., Artis, D. R., Herlyn, M., and Bollag, G. 2008. Discovery of a selective inhibitor of oncogenic B-Raf kinase with potent antimelanoma activity. Proc Natl Acad Sci USA, 105(8):3041–6.

Valencia-Sanchez, M. A., Liu, J., Hannon, G. J., and Parker, R. 2006. Control of translation and mRNA degradation by miRNAs and siRNAs. Genes Dev, 20(5): 515–24.

van der Meer, D. L. M., Degenhardt, T., Väisänen, S., de Groot, P. J., Heinäniemi, M., de Vries, S. C., Müller, M., Carlberg, C., and Kersten, S. 2010. Profiling of promoter occupancy by PPARalpha in human hepatoma cells via ChIP-chip analysis. Nucleic Acids Res, 38(9):2839–50.

van Dijk, E. L., Auger, H., Jaszczyszyn, Y., and Thermes, C. 2014. Ten years of next-generation sequencing technology. Trends Genet, 30(9):418–26.

van Helden, A. 1977. The Invention of the Telescope. Transactions of the American Philosophical SocietyNew Series, 67(4):1–67.

Vaquerizas, J., Kummerfeld, S., Teichmann, S., and Luscombe, N. M. 2009. A census of human transcription factors: function, expression and evolution. Nature Reviews Genetics, 10(4):252–63.

Vial, E. and Marshall, C. J. 2003. Elevated ERK-MAP kinase activity protects the FOS family member FRA-1 against proteasomal degradation in colon carcinoma cells. J Cell Sci, 116(Pt 24):4957–63.

Vogelstein, B., Papadopoulos, N., Velculescu, V. E., Zhou, S., Diaz, L. A., and Kinzler, K. W. 2013. Cancer genome landscapes. Science, 339(6127):1546–58.

Walther, A., Johnstone, E., Swanton, C., Midgley, R., Tomlinson, I., and Kerr, D. 2009. Genetic prognostic and predictive markers in colorectal cancer. Nat Rev Cancer, 9(7):489–99.

Warburg, O. 1925. Über den Stoffwechsel der Carcinomzelle. Journal of Molecular Medicine, 12(50):1131–37.

Warburg, O. 1956. On respiratory impairment in cancer cells. Science, 124(3215): 269–70.

Ward, A. F., Braun, B. S., and Shannon, K. M. 2012. Targeting oncogenic Ras signaling in hematologic malignancies. Blood, 120(17):3397–406.

Weinstein, I. B. 2002. Cancer. Addiction to oncogenes–the Achilles heal of cancer. Science, 297(5578):63–4.

Weinstein, I. B. and Joe, A. 2008. Oncogene addiction. Cancer Res, 68(9):3077–80; discussion 3080.

Whiteside, T. L. 2008. The tumor microenvironment and its role in promoting tumor growth. Oncogene, 27(45):5904–12.

Wiley, H. S. 2003. Trafficking of the ErbB receptors and its influence on signaling. Exp Cell Res, 284(1):78–88.

Wilson, T. R., Fridlyand, J., Yan, Y., Penuel, E., Burton, L., Chan, E., Peng, J., Lin, E., Wang, Y., Sosman, J., Ribas, A., Li, J., Moffat, J., Sutherlin, D. P., Koeppen, H., Merchant, M., Neve, R., and Settleman, J. 2012. Widespread potential for growth-factor-driven resistance to anticancer kinase inhibitors. Nature, 487(7408):505–9.

Witzel, F., Fritsche-Guenther, R., Lehmann, N., Sieber, A., and Blüthgen, N. 2015. Analysis of impedance-based cellular growth assays. Bioinformatics (Oxford, England), 31(16):btv216–2712.

Wu, X. and Brewer, G. 2012. The regulation of mRNA stability in mammalian cells: 2.0. Gene, 500(1):10–21.

Xu, T.-R., Vyshemirsky, V., Gormand, A., von Kriegsheim, A., Girolami, M., Baillie, G. S., Ketley, D., Dunlop, A. J., Milligan, G., Houslay, M. D., and Kolch, W. 2010. Inferring signaling pathway topologies from multiple perturbation measurements of specific biochemical species. Sci Signal, 3(113):ra20.

Yang, E., van Nimwegen, E., Zavolan, M., Rajewsky, N., Schroeder, M., Magnasco, M., and Darnell, J. E. 2003. Decay rates of human mRNAs: correlation with functional characteristics and sequence attributes. Genome Res, 13(8):1863–72.

Yi, T. M., Huang, Y., Simon, M. I., and Doyle, J. 2000. Robust perfect adaptation in bacterial chemotaxis through integral feedback control. Proc Natl Acad Sci USA, 97(9):4649–53.

Yoon, Y.-K., Kim, H.-P., Han, S.-W., Hur, H.-S., Oh, D. Y., Im, S.-A., Bang, Y.-J., and Kim, T.-Y. 2009. Combination of EGFR and MEK1/2 inhibitor shows synergistic effects by suppressing EGFR/HER3-dependent AKT activation in human gastric cancer cells. Mol Cancer Ther, 8(9):2526–36.

Yoon, Y.-K., Kim, H.-P., Song, S.-H., Han, S.-W., Oh, D. Y., Im, S.-A., Bang, Y.-J., and Kim, T.-Y. 2012. Down-regulation of mitogen-inducible gene 6, a negative regulator of EGFR, enhances resistance to MEK inhibition in KRAS mutant cancer cells. Cancer Lett, 316(1):77–84.

Yu, C. F., Liu, Z.-X., and Cantley, L. G. 2002. ERK negatively regulates the epidermal growth factor-mediated interaction of Gab1 and the phosphatidylinositol 3-kinase. J Biol Chem, 277(22):19382–8.

Zhang, H.-M., Chen, H., Liu, W., Liu, H., Gong, J., Wang, H., and Guo, A.-Y. 2012. AnimalTFDB: a comprehensive animal transcription factor database. Nucleic Acids Res, 40(Database issue):D144–9.

Zimmermann, G., Papke, B., Ismail, S., Vartak, N., Chandra, A., Hoffmann, M., Hahn, S. A., Triola, G., Wittinghofer, A., Bastiaens, P. I. H., and Waldmann, H. 2013. Small molecule inhibition of the KRAS-PDEδ interaction impairs oncogenic KRAS signalling. Nature, 497(7451):638–42.

Danksagungen

Zur Fertigstellung dieser Arbeit war die Unterstützung und Hilfe vieler Personen vonnöten, deren Einsatz und Einfluss an dieser Stelle gewürdigt werden soll und denen ich hiermit meinen Dank aussprechen möchte.

Anfangen möchte ich bei meinem Großvater Wolfgang Klinger, der, durch seine unermüdlichen Bemühungen meine Neugier mit aufschlussreichen Erklärungen zu stillen, es schaffte das Feuer für das Unbekannte in mir nie erlöschen zu lassen. Ich wünschte Du hättest die Fertigstellung der Frucht deines Lebenswerkes noch miterleben können.

Größten Dank gebührt auch meinen Eltern, Gert und Petra Klinger, die mit mir einige Hochs und Tiefs durchlebt haben. Euer unerschütterliches Vertrauen und unbedingte Unterstützung hat mir viel Kraft gegeben.

Darüber hinaus möchte ich meiner gesamten Arbeitsgruppe danken, deren produktive Arbeitsumgebung mit vielseitigem Profil mir viele Inspirationen bescherte. Ganz besonders gilt mein Dank hierbei Nils Blüthgen, eine kompetentere Betreuung erscheint mir unmöglich. In jedwedes Thema konntest Du dich in Windeseile hineindenken und hattest stets veritable Ratschläge zur Hand.

Ein Extra-Dank haben sich auch unsere Laboranten, Anja Sieber, Nadine Lehmann und Raphaela Fritsche verdient. Es hat Spaß gemacht mit Euch unzählige Stunden Experimente zu planen, auszuwerten und wieder neu zu planen.

Meinen herzlichsten Dank gilt den fleißigen Korrekturlesern Theresia Gutmann, Johannes Meisig, Manuela Benary, Pascal Schulthess und Michael Fähling. Ihr habt so einige Schnitzer entdeckt, die sonst unentdeckt geblieben wären.

Ich danke allen Kollegen, die an den in der Arbeit enthaltenen Studien beteiligt waren. Ohne Euren Eifer hätte keines dieser Projekte das Licht der Welt erblickt.

Zu guter Letzt möchte ich meiner liebsten Frau Mi-Kyeong und meinen bezaubernden Kindern Maya, Elias, Sarah und Luna für alles, was sie mir bedeuten, danken. Mit Euch war die Durchführung dieser Arbeit kein leichtes Unterfangen - ohne Euch aber wäre es sinnlos gewesen.